STEM CELL BIOLOGY
Development and Plasticity

ANNALS OF THE NEW YORK ACADEMY OF SCIENCES

Volume 1049

STEM CELL BIOLOGY

Development and Plasticity

Edited by Jitka Ourednick, Václav Ourednik, Donald S. Sakaguchi, and Marit Nilsen-Hamilton

The New York Academy of Sciences
New York, New York
2005

Library of Congress Cataloging-in-Publication Data

Growth Factor and Signal Transduction Conference (13th : 2004 :
 Ames, Iowa)
  Stem cell biology : development and plasticity / edited by Jitka
Ourednick ... [et al.].
   p. ; cm. — (Annals of the New York Academy of Sciences ; v.
1049)
  Includes bibliographical references and index.
  ISBN 1-57331-533-8 (cloth : alk. paper) — ISBN 1-57331-534-6
(pbk. : alk. paper)
   1. Stem cells—Congresses. 2. Embryonic stem cells—Congresses.
3. Neuroplasticity—Congresses. 4. Developmental neurobiology
—Congresses. I. Ourednick, Jitka. II. Title. III. Series.
   [DNLM: 1. Stem Cells—cytology—Congresses. 2. Stem Cells
—physiology—Congresses. 3. Cell Physiology—Congresses.
4. Neuronal Plasticity—physiology—Congresses. 5. Organogenesis
—physiology—Congresses.   W1 AN626YL v.1049 2005 / QU 325
G884 2005]
  Q11.N5 vol. 1049
  [QH588.S83]
  500 s—dc22
  [616'.02774]

                                            2005012261

GYAT/PCP

ISBN 1-57331-533-8 (cloth)
ISBN 1-57331-534-6 (paper)
ISSN 0077-8923

ANNALS OF THE NEW YORK ACADEMY OF SCIENCES
Volume 1049
May 2005

STEM CELL BIOLOGY

Development and Plasticity

Editors
JITKA OUREDNIK, VÁCLAV OUREDNIK, DONALD S. SAKAGUCHI,
AND MARIT NILSEN-HAMILTON

Conference Organizers
MARIT NILSEN-HAMILTON, DONALD S. SAKAGUCHI, JITKA OUREDNIK,
EVAN Y. SNYDER, AND VÁCLAV OUREDNIK

This volume reports the proceedings of the **Thirteenth Annual Growth Factor and Signal Transduction Conference** entitled **Stem Cell Biology: Development and Plasticity,** sponsored by the Department of Biochemistry, Biophysics and Molecular Biology, Iowa State University, and held on September 16–19, 2004 in Ames, Iowa.

CONTENTS

Financial assistance was received from:

- MOLECULAR EXPRESS, INC.
- NATIONAL INSTITUTES OF HEALTH
- OFFICE OF BIOTECHNOLOGY, IOWA STATE UNIVERSITY

Participants at the Thirteenth GFST Symposium on Stem Cell Biology in front of the Scheman Building, University of Iowa, Ames, Iowa, September 17, 2004.

Preface

*Scientific work must not be considered from the
point of view of the direct usefulness of it. It must
be done for itself, for the beauty of science, and
then there is always the chance that a scientific
discovery may become like the radium a benefit for
humanity.*

—MARIE CURIE-SKLODOWSKA
Lecture at Vassar College, May 14, 1921

*Deceptions of the senses are the truth of
perception.*

—JAN EVANGELISTA PURKYNĚ (PURKINJE)

This volume of the *Annals* entitled *Stem Cell Biology: Development and Plasticity*
embodies many of the papers presented at the thirteenth annual Growth Factor and
Signal Transduction Conference (GFST), held September 16–19, 2004 at the
Scheman Building on the Iowa State University campus in Ames, Iowa. As has been
true since the inception of the GFST Symposia series in 1988, the organizers invited
internationally renowned investigators to present their work and to consider scientific
issues of mutual concern.

ﾞﾑ

Stem cells have garnered tremendous interest recently in the scientific, clinical
and public arenas. This symposium on the biology of stem cells comes at a time of
considerable excitement in the areas of developmental biology, biotechnology, and
biomedical sciences. Not only do stem cells provide unique opportunities for major
advances in our understanding of developmental processes, but also, by means of
isolating, maintaining, and manipulating embryonic and organotypic stem cells in
culture, unprecedented therapeutic possibilities are afforded for repair of damaged
or diseased organs and tissues.

The past decade has seen enormous progress in our understanding of the specific
requirements of stem cells and of their potential plasticity, not only in the context of
the central nervous system, but also in other tissues as well. A plethora of growth
factors and substrates were discovered and shown *in vitro* to improve and guide proliferation, migration, and differentiation of stem cells into specified lineages. However, the behavior of stem cells in tissue culture does not reflect fully how the same
cells will act when grafted into the living host, where multiple factors (both known
and not discovered yet) converge to influence biological processes. It is true that
both intrinsic (intracellular) and extrinsic (environmental) factors govern the reciprocal interaction of cells, instructing them to adopt specific cellular fates. The microenvironment encountered by the transferred stem cells may vary with respect to
age, injury, deprivation, or other circumstances, which will affect the survival and
incorporation of the donor cells. On the other hand, transplanted stem cells can act

Ann. N.Y. Acad. Sci. 1049: ix–xii (2005). © 2005 New York Academy of Sciences.
doi: 10.1196/annals.1334.020

to influence and change host cells in their vicinity. They may release agents that alter the activity or resilience of the host tissue that is already damaged or in danger. These dynamic interactions occur naturally and respond to each other in time and space. Teasing out and understanding these interactions poses a major challenge that must be faced in order to develop realistic cell therapies and enhance normal tissue resistance and regeneration.

As work in the complex field of stem cell biology has become more and more diversified and specialized and we all have begun to work in our own small enclaves, the need to disseminate knowledge across disciplines becomes indisputable. The theme of the conference was reflected in six topics[a] introduced by plenary lectures: *Regulators of Stem Cell Biology* (Theo Palmer and Derek van der Kooy) discussed molecular mechanisms defining and modulating the very early stages in stem cell differentiation. *Regional Determinants of Cell Fate* (Marie Csete and Steven Goldman) described the influence of oxidative stress on stem cell biology *in vitro* and of the environment on gliogenesis. *Intrinsic Determinants of Stem Cells* (Elena Cattaneo, Ernest Arenas, and Linheng Li) covered the topic of environmental factors influencing specific signaling pathways in the differentiation of stem cells into specialized mature cell types. *Transdifferentiation* (Paul Sanberg and Carl Gregory) illustrated the capacity of stem cells to switch between different germ layer commitments. *Stem Cell Responses to Perturbation* (Donald Sakaguchi, Eva Syková, Clive Svendsen) portrayed the behavior and therapeutic potential of stem cells grafted into developing and impaired postnatal central nervous system. *Reciprocal Communication between Graft and Host* (Jitka Ourednik and Evan Snyder) concluded the plenary lectures by emphasizing the importance of the dialogue beween grafted cells and their host environment in maintenance of tissue homeostasis and its importance in evoking reparative mechanisms in the compromised host central nervous syustem.

The GFST Conferences have a long and cherished history at Iowa State University. An important purpose is to promote new collaboration and interdisciplinary interactions between individuals working in specialized areas who might not have encountered each other in less-focused meetings. By doing so, participants, ranging from distinguished scientists to highly promising students and postdoctoral fellows meet in an informal atmosphere that allows for extensive discussions both during and between scientific sessions.

Many people were responsible for assisting us, the organizers, in making the symposium itself a success and in preparing the papers for this resulting *Annals* volume. First and foremost, of course, we owe our thanks to the conference presenters and the contributors to this volume. Second, the graduate and undergraduate students at Iowa State University deserve thanks for attending to many of the administrative and technical details of the conference. Third, we are grateful to Rohini Ramaswamy, Gulshan Singh, and Allison Rudig, the Symposium Office, for their devoted assistance in the logistical aspects of the symposium. Without these individuals neither the symposium itself nor this book would exist. Finally, we would like to acknowledge the joint financial support for the meeting from the National Institutes of Health; the Office of Biotechnology, the Graduate College, and

[a]Regrettably, the papers of Drs. Palmer, van der Kooy, Goldman, and Snyder are not included in this volume.

the Department of Biochemistry, Biophysics and Molecular Biology of Iowa State University; Merck & Co.; and Molecular Express Inc.

—JITKA OUREDNIK
—VÁCLAV OUREDNIK
—DONALD SAKAGUCHI
—MARIT NILSEN-HAMILTON

ॐ

As a final note, we offer the following personal dedications to those persons whose sustenance and love has allowed us to undertake the work we have collected and reported in this volume:

To my mother Květa and father Stanislav Hanzík for igniting my interest in nature and to my grandmothers Karlička and Mánička for their warm care. — *Jitka*

To my parents Maruška and Jiří Ourednik for their love and continuous encouragement. — *Václav*

To my parents Mary and Ed Sakaguchi, and to my wife Kathleen Ary, for their support throughout my scientific career; to my current and former students for their intellectual curiosity and to the late Christopher Reeve for his efforts in support of stem cell research. — *Donald*

To my parents Joan and Tallak Nilsen, and to my husband Richard T. Hamilton, for their steadfast love and support over the years. — *Marit*

Oxygen in the Cultivation of Stem Cells

MARIE CSETE

Emory Anesthesiology Research Laboratories, Atlanta, Georgia 30322, USA

ABSTRACT: Cultivation of stem cells, like all cells in culture, is performed under conditions that cannot and do not replicate normal physiologic conditions.
For example, direct exposure of cultured monolayer cells to serum contents is
normally prevented *in vivo* by the vasculature. The heterogeneity of cells and
signals between different cell types in an organ is certainly not captured when
a single cell type is grown and studied *in vitro*. Gases, in particular, are not accounted for in routine tissue culture. Oxygen is fundamental for life and its
concentration is an important signal for virtually all cellular processes. Nonetheless, oxygen is rarely taken into account in culturing stem and other cells.
This review will summarize work that highlights the importance of considering
oxygen conditions for culturing and manipulating stem cells. Emphasis is
placed on major phenotypic changes in response to oxygen, recognizing that
oxygen-mediated transcriptional and post-translational effects are enormously
complex, and beyond the scope of this review. The review emphasizes that
oxygen is an important signal in all major aspects of stem cell biology including
proliferation and tumorigenesis, cell death and differentiation, self-renewal,
and migration.

KEYWORDS: oxygen; reactive oxygen species; self-renewal; stem cell

GASES IN AIR VS. TISSUES

Carbon dioxide (despite global warming) is present in parts per million in air. The
traditional 5–7% CO_2 used in tissue culture serves as a buffer, and also reflects levels *in vivo*. Normal arterial blood contains CO_2 of 40 mmHg or 5.3%. In room air
21% of the gas is oxygen, and in traditional incubators supplied by room air and
buffered with 5% CO_2, the concentration of oxygen is 20% (and nitrogen is 75%).
However, at the tissue level, oxygen concentrations *in vivo* are significantly less. Arterial blood is about 12% oxygen, and the mean tissue level of oxygen is about 3%,
with considerable local and regional variation. These means are for adult organs and
tissues. Mean oxygen tension at the tissue level in embryos (where stem cells are enriched relative to adult tissues) is considerably less. Although many papers refer to
the low oxygen levels in embryos as "hypoxic," the low oxygen levels are actually
"normoxic" for the time and place in development.

Clearly, controlling oxygen in tissue culture is not easy (room air is convenient),
and tissue culture has yielded remarkable insights despite the discrepancy between

Address for correspondence: Marie Csete, M.D., Ph.D., Emory Anesthesiology Research
Labs, 1462 Clifton Rd. N.E., Room 420, Atlanta, GA 30322. Voice: 404-712-2588; fax: 404-
712-2585.
Marie_Csete@emoryhealthcare.org

Ann. N.Y. Acad. Sci. 1049: 1–8 (2005). © 2005 New York Academy of Sciences.
doi: 10.1196/annals.1334.001

oxygen concentrations in incubators vs. *in vivo*. Nonetheless, more oxygen is not necessarily better in the case of tissue culture. Clinicians have known for many decades that exposure of cells to high levels of oxygen is toxic, particularly in the lung. Furthermore, the importance of oxygen concentration control is highlighted by the enormous, intricate mechanisms used to maintain oxygen homeostasis in the body. Multiple physiologic feedback loops at the cellular and inter-organ levels maintain oxygen by varying minute ventilation, cardiac output, vascular tone and density, red cell mass, and energy expenditure. At the cellular level, changes in oxygen availability are sensed and survival responses initiated when oxygen is limited, coordinated largely through stabilization of hypoxic inducible factor 1α (HIF-1α).[1] Independent of HIF-1α, changes in oxygen induce diverse transcriptional responses, post-translational modifications, and changes in subcellular protein trafficking.

PRACTICAL ISSUES IN CONTROLLING OXYGEN IN THE LAB

Current technologies do not allow the exquisite control over oxygen *in vitro* that is accomplished *in vivo*. First, multiple oxygen sensors embedded at various levels in the tissue culture would be necessary and are not standard technologies, though novel sensors are being developed that could improve the sensing[2] and control of oxygen in culture. But some basics need to be pointed out. Incubators are now routinely available that allow control over oxygen levels during cultivation of cells. However, use of these incubators does not prevent reperfusion of room air (reperfusion injury) when the cells are removed for medium changes, passaging, or other manipulations. In order to have relatively constant levels of controlled oxygen, the cells must be manipulated in a workstation in which oxygen and CO_2 are also controllable. In addition, the level of oxygen that the incubators (or closed workstations) deliver is not the same as that seen at the level of the tissue culture monolayer. The oxygen tension at the (tissue culture) cell level depends on the depth of medium, density of cells, and cellular respiration, in addition to the oxygen level in the incubator[3] and is very difficult to control precisely. Tissue engineers have generally been more attentive than biologists in accounting for oxygen as a major variable in *ex vivo* cell cultivation.[4] Nonetheless, it is possible (without the use of bioreactors) to lower oxygen tension in tissue culture for prolonged periods of time to levels that are closer to normal physiologic levels than in traditional tissue culture with 20% oxygen, without subjecting cells to hypoxic stress.[5] Under these conditions, the phenotype of cultured stem cells is dramatically altered.

ALL ASPECTS OF STEM CELL BIOLOGY ARE ALTERED BY CHANGING OXYGEN IN CULTURE

Stem cells can be defined as cells having high proliferative potential with the ability to self-renew. They are undifferentiated, but can generate daughters of more than one distinct phenotype, and stem cells also migrate to areas of injury.[6] All these features fundamental to stem cell function are dramatically altered by oxygen in tissue culture.

PROLIFERATION OF STEM CELLS IS ENHANCED IN LOWER, PHYSIOLOGIC OXYGEN CULTURE CONDITIONS

Virtually all stem cells cultivated in lower, physiologic oxygen (not frankly hypoxic conditions) proliferate more than in traditional 20% O_2 environments. Formal studies on enhanced proliferation in lower oxygen have been reported for (rat) CNS-derived multipotent stem cells,[7] (rat) fetal-derived neural crest stem cells,[8] adult murine skeletal muscle satellite cells,[9] (rat) marrow-derived mesenchymal stem cells[10] and in CD34+ marrow progenitor populations.[11] The practical benefit of low oxygen culture, then, is that relatively rare cell populations are more easily expanded *in vitro*. Proliferation of first-trimester trophoblast cells is also increased in 2% oxygen (vs. 20%) in a bioreactor system.[12]

In the case of CD34+ progenitor cells, enhanced proliferation in low oxygen does not occur at the expense of progenitor cell renewal.[13] Low oxygen also supports maintenance of human cord blood progenitors in culture.[14] Whether low oxygen culture conditions can, in general, help maintain stem cell self-renewal *in vitro* is not known. But this area deserves investigation, particularly in light of reports that antioxidant treatment of hematopoietic cells enhances self renewal, acting via the ataxia telangiectasia mutated (ATM) gene.[15] In this regard, reactive oxygen species were reported long ago to limit self-renewal of hematopoietic progenitor cell populations.[16]

Enhanced proliferation of stem cells in lower oxygen may in part reflect the normal relatively low O_2 environment of stem cells *in vivo*, even in adult organs. Marrow, in which oxygen tensions are lower than in peripheral blood, harbors well-defined hematopoietic stem cell populations as well as other less characterized populations. Models of marrow architecture suggest that stem cells normally reside in a particularly low oxygen niche of marrow,[17] and that oxygen may be a primary determinant of stem cell mobilization from marrow in disease processes.[18] The renal papilla also harbors a stem cell population, and this area of the kidney normally has a very low tissue-level oxygen tension.[19] Furthermore, early tissue and organ specification events in embryos are mediated by oxygen. The lowest areas of oxygen in embryos for example, induce expression of the mesodermal T box gene, brachyury and BMP4 (bone morphogenetic protein-4) setting up the early maturation of mesoderm.[20] Thus, stem cell–mediated events analogous to those in embryonic development may be masked or underestimated *in vitro* when aphysiologic high oxygen conditions are used for stem cell cultivation.

It is also important to note that proliferative senescence in general (not only in true stem cells) is delayed or halted by culture in low oxygen, and in these low oxygen conditions DNA damage is reduced.[21]

APOPTOSIS IS REDUCED IN STEM CELL POPULATIONS CULTURED IN LOWER OXYGEN CONDITIONS

Fetal rat–derived CNS stem cells undergo significantly less apoptosis in culture in low oxygen than in traditional 20% O_2 conditions.[7] Similarly, murine muscle satellite stem cells also undergo less apoptosis in 6% O_2 than in 20% O_2, and low oxygen culture conditions blunt apoptosis in satellite cells deficient in antioxidant genes (Csete, unpublished material). CD34[+] marrow progenitor cells cultured in 5% oxygen undergo less (or delayed) apoptosis compared to those cultured in 20% O_2,[22] and similar results have been reported for a variety of differentiated cell types.[23] Interestingly, oxygen-induced apoptosis may be a prominent mechanism for clearing fetal nucleated red cells from the maternal circulation,[24] as the maternal circulation has a significantly higher oxygen content than the fetal circulation.

The anti-apoptotic effect of low oxygen may be a reflection of the fact that fewer reactive oxygen species are generated in culture in low vs. high oxygen conditions. In primary murine myoblast cultures, about 1/3 fewer reactive oxygen species are generated in 6% vs. 20% oxygen cultures (S. Lee and M. Csete, in preparation). Other investigators, using similar redox reactive dyes, have shown similar reduction in reactive oxygen species when oxygen is lowered in culture.[25] Taken together with an independent effect on proliferation, the reduction in apoptosis in low oxygen can increase yield of cultured stem cells significantly.[7]

DIFFERENTIATION PATTERNS OF STEM CELLS ARE ALSO AFFECTED BY LOWER OXYGEN CONDITIONS IN CULTURE

Mesoderm-derived stem cells show remarkable differentiation changes in response to lower oxygen in culture. Both skeletal muscle stem cells and commonly used mesenchymal stem cell lines are more likely to undergo myogenesis rather than adipogenesis in lower oxygen culture than in 20% O_2.[9] In very early murine satellite cell activation, low oxygen is associated with increased likelihood that the cells will express multiple myogenic basic-loop-helix genes, core regulators of myogenesis.[9] But the effect of oxygen is likely to be much more complex. Tumor necrosis factor-alpha, for example, is induced by reactive oxygen species, and is a potent inhibitor of skeletal muscle differentiation (reviewed in Reid and Li[26]). In addition, the core regulator of adipogenesis, PPAR-gamma, upregulated in high oxygen conditions,[9] can actively repress MyoD promoter activity.[27] This observation is particularly important since it suggests that there are oxygen-responsive mechanisms that can both actively promote differentiation of one possible lineage (adipocytes) and repress another possible lineage (muscle). HIF-1α may be involved in the suppression of adipogenesis in hypoxic conditions via DEC1/Stra13-mediated repression of PPAR-gamma.[28] Cardiac muscle differentiation is also impaired by reactive oxygen species: During *in vitro* cardiac differentiation from embryonic stem cells, increased reactive oxygen species decrease expression of mef2c via Rac.[29]

A direct connection to stemcell–mediated phenotypic changes with aging is suggested by the differentiation patterns described above. Aging is associated with increased reactive oxygen species stress, and with accumulation of fat in skeletal muscle,[30] marrow,[31] and liver.[32] In addition aging is associated with loss of muscle mass. Taken together the evidence presented above suggests that muscle stem cells in an aged environment characterized by oxidant stress are likely to have a low threshold for adipogenesis and an increased threshold for myogenesis and so may contribute the altered phenotype of skeletal muscle with aging.

As noted above, experiments in which oxygen is reduced around the cultured cells are associated with accumulation of fewer intracellular oxidants, a feature documented in many types of cells using indirect measures that rely on fluorescent dyes activated by oxidative changes. Thus, many of the changes attributed to oxygen in culture are likely due to changes in a broad range of reactive oxygen species. In the case of skeletal muscle stem cells, ongoing work in our group suggests that deletion of antioxidant pathways strongly inhibits terminal differentiation of skeletal muscle (not just the initial fate choice between muscle and another pathway). But quantitative and qualitative details of the antioxidant/oxidant balance are critical, and overexpression of antioxidants can also be deleterious. For example, some mice strains overexpressing the cytosolic form of superoxide dismutase (CuZnSOD) develop myopathies (whereas others do not), pointing to the fine line between protection and toxicity in oxidant/antioxidant activity.[33]

Central and peripheral nervous system stem cells in culture also show remarkable changes in differentiation patterns when exposed to low oxygen in culture. For CNS stem cells in low oxygen, differentiation is skewed toward a dopaminergic neuron phenotype, and serotonergic neuron differentiation is also enhanced.[7] (It is important to note that mean tissue level oxygen in the brain across mammalian species is about 1.5%.) Neural crest stem cells cultured in lower oxygen conditions show a similar pattern with enhanced differentiation of sympathoadrenal neurons. In fact, low oxygen unmasked this phenotype in culture, as 20% culture conditions were not conducive to differentiation of this lineage.[8]

Very little work in culture has been done with neurons at low, non-hypoxic oxygen conditions, with the majority of work focusing on truly hypoxic, oxygen-deprived conditions (mimicking stroke in a dish) as the experimental model. Usually, in tissue culture experiments of ischemic neurons or ischemic-reperfusion injury of neurons, the control conditions are 20% oxygen. Comparison of true hypoxic conditions is not made to more normal physiologic oxygen conditions. And on the other extreme when oxidative stress is being studied in neurons, oxidants are usually added above and beyond the already hyperoxic 20% oxygen conditions, again the control not reflecting physiologic norms. The (negative) impact of ignoring physiologic oxygen in neuronal differentiation, then, is largely unexplored. Although there is a deep understanding of hypoxic responses (especially those mediated via HIF-1α), the expression level changes induced at the extremes of normoxia are also likely very important in development and have not been at all well-studied. One feature of development that is clearly missed by usual tissue culture environment is the feedback between parallel developing blood vessels and brain parenchyma, much of it revolving around oxygen sensing of changes within normal physiologic oxygen levels.[34,35]

STEM CELLS, MIGRATION, AND OXYGEN

In order to perform their regenerative function, stem cells must be able to migrate.[6] Marrow-derived cells can be found in a variety of organs after marrow transplantation[36–38] and pharmacologic tools to enhance mobilization of stem cell populations are under intensive investigation because of their presumed importance in translational stem cell therapies. Among the growth factors known to enhance mobilization of some marrow stem cells into the circulation are VEGF[39] and erythropoietin, both classically oxygen-regulated.[40] Since injuries that induce marrow mobilization often involve hypoxia or oxidative stress, oxygen is likely fundamental to this aspect of stem cell biology as well. A recent report shows that mobilization of progenitor populations is induced by stromal cell–derived factor (SDF-1) expression (under control of HIF-1α) in ischemic tissues, generating gradients that guide trafficking.[41]

The regulation of stem cell migration and the regulation of tumor cell migration have overlapping mechanisms. SDF-1 is a CXC cytokine also involved in metastasis of a variety of tumors.[42,43] Matrix metalloproteases and hepatocyte growth factor (also prominently oxygen-regulated) are among many overlapping mechanisms that control both normal marrow progenitor mobilization[44] and tumor metastasis.[45,46] The hypoxic environment in tumors mediates upregulation of genes that enhance tumor survival (including erythropoietin and VEGF) and radiation resistance.[47] Thus the hypoxic core of tumors may be a source not only of signals for metastasis, but also protection of cancer stem cells. In addition chromosomal instability is related directly to oxidant stress[48] and the inability to protect cultured human embryonic stem cells from chromosomal instability is currently a major problem in their cultivation.[49]

In summary, changes in oxygen surrounding cultured cells induce diverse transcriptional and protein-level responses reflective of responses to oxygen *in vivo*. Some of these oxygen-mediated changes can be harnessed to advantage in tissue culture, and used to enhance stem cell expansion *in vitro*, or to promote particular patterns of differentiation for study. Hypoxic conditions and hyperoxic, oxidative conditions in culture can also be used to mimic disease states in stem cell populations. Oxygen is not the only physiologic gas with important effects on stem cell populations. Although nitrogen is considered inert in this regard, both nitric oxide (NO) and carbon monoxide (CO) also induce important changes in cultured cells. In order to understand the control of stem cells *in vitro*, further characterization of the optimal gaseous microenvironment is essential.

REFERENCES

1. IYER, N.V., L.E. KOTCH, F. AGANI, *et al.* 1998. Cellular and developmental control of O_2 homeostasis by hypoxia-inducible factor 1 alpha. Genes Dev. **12:** 149–162.
2. CHANG-YEN, D.A., & B.K. GALE. 2003. An integrated optical oxygen sensor fabricated using rapid-prototyping techniques. Roy. Soc. Chem. **3:** 297–301.
3. TOKUDA, Y., S. CRANE, Y. YAMAGUCHI, *et al.* 2000. The levels and kinetics of oxygen tension detectable at the surface of human dermal fibroblast cultures. J. Cell Physiol. **182:** 414–420.
4. MUSCHLER, G.F., C. NAKAMOTO, & L.G. GRIFFITH. 2004. Engineering principles of clinical cell-based tissue engineering. J. Bone Joint Surg. Am. **86-A:** 1541–1558.
5. STORCH, A., G. PAUL, M. CSETE, *et al.* 2001. Long-term proliferation and dopaminergic differentiation of human mesencephalic neural precursor cells. Exp. Neurol. **170:** 317–25.
6. BLAU, H.M., T.R. BRAZELTON, & J.M. WEIMANN. 2001. The evolving concept of a stem cell: entity or function? Cell **105:** 829–841.
7. STUDER, L., M. CSETE, S.H. LEE, *et al.* 2000. Enhanced proliferation, survival, and dopaminergic differentiation of CNS precursors in lowered oxygen. J. Neurosci. **20:** 7377–7383.
8. MORRISON, S.J., M. CSETE, A.K. GROVES, *et al.* 2000. Culture in reduced levels of oxygen promotes clonogenic sympathoadrenal differentiation by isolated neural crest stem cells. J. Neurosci. **20:** 7370–7376.

9. CSETE, M., J. WALIKONIS, N. SLAWNY, et al. 2001. Oxygen-meidated regulation of skeletal muscle satellite cell proliferation and adipogenesis in culture. J. Cell Physiol. **189:** 189–196.
10. LENNON, D.P., J.M. EDMISON & A.I. CAPLAN. 2001. Cultivation of rat marrow-derived mesenchynmal stem cells in reduced oxygen tension: effects on in vitro and in vivo osteochondrogenesis. J. Cell Physiol. **187:** 345–55.
11. REYKDAL, S., C. ABBOUD & J. LIESVELD. 1999. Effect of nitric oxide production and oxygen tension on progenitor preservation in ex vivo culture. Exp. Hematol. **27:** 441–450.
12. MA, T., S.T. YANG, & D.A. KNISS. 2001. Oxygen tension influences proliferation and differentiation in a tissue-engineered model of placental trophoblast-like cells. Tissue Eng. **7:** 495–506.
13. IVANOVIC, Z., P. DELLO SBARBA, F. TRIMOREAU, et al. 2000. Primitive human HPCs are better maintained and expanded at 1 percent oxygen than at 20 percent. Transfusion **40:** 1482–1488.
14. KOLLER, M.R., J.G. BENDER, E.T. PAPOUTSAKIS, & W.M. MILLER. 1992. Effects of synergistic cytokine combinations, low oxygen, and irradiated stroma on the expansion of human cord blood progenitors. Blood **80:** 403–411.
15. ITO, K., A. HIRAO, F. ARAI, et al. 2004. Regulation of oxidative stress by ATM is required for self-renewal of haematopoietic stem cells. Nature **431:** 997–1002.
16. MEAGHER, R.C., A.J. SALVADO, & D.G. WRIGHT. 1988. An analysis of the multilineage production of human hematopoietic progenitors in long-term bone marrow culture: evidence that reactive oxygen intermediates derived from mature phagocytic cells have a role in limiting progenitor cell self-renewal. Blood **72:** 273–281.
17. CHOW, D.C., L.A. WENNING, W.M. MILLER, & E.T. PAPOUTSAKIS. 2001. Modeling pO(2) distributions in bone marrow hematopoietic compartment. I. Krogh's model. Biophys. J. **81:** 675–684.
18. ANTONIOU, E.S., S. SUND, E.N. HOMSI, et al. 2004. A theoretical simulation of hematopoietic stem cells during oxygen fluctuations: prediction of bone marrow responses during hemorrhagic shock. Shock **22:** 415–422.
19. OLIVER, J.A., O. MAAROUF, F.H. CHEEMA, et al. 2004. The renal papilla is a niche for adult kidney stem cells. J. Clin. Invest. **114:** 795–804.
20. RAMIREZ-BERGERON, D.L., A. RUNGE, K.D. DAHL, et al. 2004. Hypoxia affects mesoderm and enhances hemangioblast specification during early development. Development **131:** 4623–4634.
21. PARRINELLO, S., E. SAMPER, A. KRTOLICA, et al. 2003. Oxygen sensitivity severely limits the replicative lifespan of murine fibroblasts. Nat. Cell Biol. **5:** 741–747.
22. MOSTAFA, S.S., W.M. MILLER, & E.T. PAPOUTSAKIS. 2002. Oxygen tension influences the differentiation, maturation and apoptosis of human megakaryocytes. Br. J. Haematol. **111:** 879–889.
23. ERKKILA, K., V. PENTIKAINEN, M. WIKSTROM, et al. 1999. Partial oxygen pressure and mitochondrial permeability transition affect germ cell apoptosis in the human testis. J. Clin. Endocrinol. Metab. **84:** 4253–4259.
24. KONDO, T., A. SEKIZAWA, H. SAITO, et al. 2002. Fate of fetal nucleated erythrocytes circulating in maternal blood: Apoptosis is induced by maternal oxygen concentration. Clin. Chem. **48:** 1618–1620.
25. PFAU, J.C., J.C. SCHNEIDER, A.J. ARCHER, et al. 2004. Environmental oxygen tension affects phenotype in cultured bone marrow-derived macrophages. Am. J. Physiol. Lung Cell Mol. Physiol. **286:** L354–362.
26. REID, M.B., & Y.P. LI. 2001. Tumor necrosis factor-alpha and muscle wasting: a cellular perspective. Respir. Res. **2:** 269–272.
27. HUNTER, J.G., M.F. VAN DELFT, R.A. RACHUBINSKI, & J.P. CAPONE. 2001. Peroxisome proliferator-activated receptor gamma ligands differentially modulate muscle cell differentiation and MyoD gene expression via peroxisome proliferator-activated receptor gamma -dependent and -independent pathways. J. Biol. Chem. **276:** 38297–38306.
28. YUN, Z., H.L. MAECKER, R.S. JOHNSON, & A.J. GIACCIA. 2002. Inhibition of PPAR gamma 2 gene expression by the HIF-1-regulated gene DEC1/Stra13: a mechanism for regulation of adipogenesis by hypoxia. Dev. Cell **2:** 331–341.

29. PUCEAT, M., P. TRAVO, M.T. QUINN, P. & P. FORT. 2003. A dual role of the GTPase Rac in cardiac differentiation of stem cells. Mol. Biol. Cell **14:** 2781–2792.
30. BEGGS, M.L., R. NAGARAJAN, J.M. TAYLOR-JONES, et al. 2004. Alterations in the TGF-beta signaling pathway in myogenic progenitors with age. Aging Cell **3:** 353–361.
31. KIRKLAND, J.L., T. TCHKONIA, T. PIRTSKHALAVA, et al. 2002. Adipogenesis and aging: does aging make fat go MAD? Exp. Gerontol. **37:** 757–767.
32. GOODPASTER, B.H. 2002. Measuring body fat distribution and content in humans. Curr. Opin. Clin. Nutr. Metab. **5:** 481–487.
33. RANDO, T.A. & EPSTEIN C.J. 1999. Copper/zinc superoxide dismutase: More is not necessarily better! Ann. Neurol. **46:** 135–136.
34. RUBERTE, J., A. CARRETERO, M. NAVARRO, et al. 2003. Morphogenesis of blood vessels in the head muscles of avian embryo: spatial, temporal, and VEGF expression analyses. Dev. Dyn. **227:** 470–483.
35. MCCARTY, J.H., A. LACY-HULBERT, A. CHAREST, et al. 2004. Selective ablation of {alpha}v integrins in the central nervous system leads to cerebral hemorrhage, seizures, axonal degeneration and premature death. Development Dec 2: Epub ahead of print.
36. GUSSONI, E., Y. SONEOKA, C.D. STRICKLAND, et al. 1999. Dystrophin expression in the mdx mouse restored by stem cell transplantation. Nature **401:** 390–394.
37. MEZEY, E., S. KEY, G. VOGELSANG, et al. 2003. Transplanted bone marrow generates new neurons in human brains. Proc. Natl. Acad. Sci. **100:** 1364–1369.
38. KALE, S., A. KARIHALOO, P.R. CLARK, et al. 2003. Bone marrow stem cells contribute to repair of the ischemically injured renal tubule. J. Clin. Invest. **112:** 42–49.
39. RABBANY, S.Y., B. HEISSIG, K. HATTORI, & S. RAFII. 2003. Molecular pathways regulating mobilization of marrow-derived stem cells for tissue revascularization. Trends Mol. Med **9:** 109–117.
40. KESSINGER, A. & J.G. SHARP. 2003. The whys and hows of hematopoietic progenitor and stem cell mobilization. Bone Marrow Transplant. **31:** 319–329.
41. CERADINI, D.J., A.R. KULKARNI, M.J. CALLAGHAN, et al. 2004. Progenitor cell trafficking is regulated by hypoxic gradients through HIF-1 induction of SDF-1. Nat. Med. **10:** 858–864.
42. MORI, T., R. DOI, M. KOIZUMI, et al. 2004. CXCR4 antagonist inhibits stromal cell-derived factor 1-induced migration and invasion of human pancreatic cancer. Mol. Cancer Ther. **3:** 29–37.
43. ZEELENBERG, I.S., L. RUULS-VAN STALLE, & E. ROOS. 2003. The chemokine receptor CXCR4 is required for outgrowth of colon carcinoma micrometastases. Cancer Res. **63:** 3833–3839.
44. KOLLET, O., S. SHIVTIEL, Y.Q. CHEN, et al. 2003. HGF, SDF-1, and MMP-9 are involved in stress-induced human CD34+ stem cell recruitment to the liver. J Clin Invest. **112:** 160–169.
45. FONG, S., Y. ITAHANA, T. SUMIDA, et al. 2003. Id-1 as a molecular target in therapy for breast cancer cell invasion and metastasis. Proc. Natl. Acad. Sci. **100:** 13543–13548.
46. JANKOWSKI, K., M. KUCIA, M. WYSOCZYNSKI, et al. 2003. Both hepatocyte growth factor (HGF) and stromal-derived factor-1 regulate the metastatic behavior of human rhabdomyosarcoma cells, but only HGF enhances their resistance to radiochemotherapy. Cancer Res. **63:** 7926–7935.
47. ACS, G., M. CHEN, X. XU, P. ACS, et al. 2004. Autocrine erythropoietin signaling inhibits hypoxia-induced apoptosis in human breast carcinoma cells. Cancer Lett. **214:** 243–251.
48. WOO, R.A. & R.Y. POON. 2004. Activated oncogenes promote and cooperate with chromosomal instability for neoplastic transformation. Genes Dev. **8:** 1317–1330.
49. DRAPER, J.S., K. SMITH, P. GOKHALE, et al. 2004. Recurrent gain of chromosomes 17q and 12 in cultured human embryonic stem cells. Nat. Biotechnol. **22:** 53–54.

DNA Damage–Induced Programmed Cell Death

Potential Roles in Germ Cell Development

YUKIKO YAMADA AND CLARK R. COFFMAN

Department of Genetics, Development and Cell Biology, Iowa State University, Ames, Iowa 50011, USA

ABSTRACT: The detection of DNA damage is necessary to protect against proliferation of potentially harmful cells and often results in cell cycle arrest and programmed cell death. Key components of DNA damage signaling networks include ATM, CHK2, p53, and Bax. Mutations in these damage signaling systems are linked to tumorigenesis and developmental abnormalities. Expression of some of these genes in primordial germ cells (PGCs) argues that PGCs may utilize DNA damage–induced signaling mechanisms to select against germ cells that are genetically defective, thus maintaining the integrity of the germline. This paper summarizes the roles of these DNA damage signaling molecules and addresses their potential involvement in germ cell development.

KEYWORDS: DNA damage; programmed cell death; primordial germ cells; ATM; p53

INTRODUCTION

In animals, genetic information is passed from one generation to the next via the germline. The formation of the germline begins when primordial germ cells (PGCs) differentiate during embryogenesis. Through a series of complex developmental processes, PGCs give rise to gametes in adults. Proper regulation of germ cell development is vital to integrated transmission of genetic material between generations.

Germ cell programmed cell death (PCD) has been shown to be essential in the prevention of germ cell tumor formation and control of germ cell numbers.[1–3] Because of its importance in human medicine, the regulation of germ cell PCD has been the subject of numerous cancer and reproductive studies. Since the DNA molecule is highly reactive and susceptible to damage by cellular metabolites, most organisms appropriate DNA damage–induced PCD to eliminate genetically flawed cells. Molecular analyses using organisms from nematodes to mammals have revealed conserved molecular components of the genome surveillance machinery. While it seems

Address for correspondence: Yukiko Yamada, Department of Genetics, Development and Cell Biology, Iowa State University, 3238 Molecular Biology Building, Ames, IA 50011-3260. Voice: 515-294-4734, fax: 515-294-6755.
    yukiko@iastate.edu

Ann. N.Y. Acad. Sci. 1049: 9–16 (2005). © 2005 New York Academy of Sciences.
doi: 10.1196/annals.1334.002

reasonable that germ cells might utilize DNA damage–induced PCD, this possibility has not been explored.

This review highlights some of the key findings in DNA damage–induced PCD. Focusing mainly on *Drosophila* and mammalian systems, we note some interesting observations and unresolved questions in this rapidly advancing field. Additionally, we comment on implications these findings may have for germ cell development.

AN OVERVIEW OF DNA DAMAGE–INDUCED
PROGRAMMED CELL DEATH

Dissection of the intricate molecular networks of DNA damage signaling has identified gatekeeper molecules that prevent defective cells from potentially harmful proliferation. The first line of defense employs DNA damage sensor molecules that monitor the state of the genome. Cells incurring DNA damage respond by halting the cell cycle to allow repair or promoting PCD if the damage is beyond repair.[4] Failure to respond properly to DNA damage can result in the loss of genomic material and carcinogenesis.[5–8]

Gatekeeper molecules in the cellular response to DNA damage include ataxia telangiectasia mutated (ATM) and CHK2 molecules. In addition, p53 is a key regulator of DNA damage–induced PCD. ATM-mediated damage signaling pathways have been linked to p53 activation. Activated p53 acts primarily as a transcription factor, entering the nucleus to bind to target enhancers. Numerous downstream pro-PCD genes are regulated by p53, including mammalian *Bax* and *Drosophila reaper* (*rpr*), *hid*, and *sickle* (*skl*) (FIG. 1).

ATM

Eukaryotic genome integrity is maintained in part by cell cycle checkpoint pathways. One of these involves the activation of a phosphatidylinositol-3 kinase-related kinase, ATM, when DNA double-strand breaks (DSBs) are recognized.[9] ATM triggers a biochemical cascade that relays and amplifies the damage signal to downstream effectors.[10] After the induction of DSBs by ionizing radiation (IR), ATM dimers autophosphorylate to become potent protein kinase monomers that carry out the signaling function.[11] Simultaneously, a fraction of ATM is also found adhered to the site of DNA damage.[12] To date, the mechanisms by which ATM is recruited to the DSB sites and how DSBs trigger ATM autophoshprylation remain unclear.

In humans, *ATM* mutations are present in persons with ataxia-telangiectasia (A-T), leading to progressive neuromotor dysfunction, immunodeficiency, infertility, and higher susceptibility to cancer.[13,14] Notably, the phenotypes of *ATM* knockout mice resemble those of A-T patients.[15,16] The importance of ATM in DNA damage–induced PCD signaling is exemplified by observations of *ATM*-deficient mice in response to IR. In wild-type mice, IR leads to PCD in many regions of the CNS, whereas IR fails to induce cell death in *ATM*-deficient mice.[17]

Drosophila lacking the ATM ortholog, *dATM*, die as pupae or manifest eye and wing malformations.[18] Flies carrying hypomorphic mutations in *dATM* have small, rough eyes and bristle abnormalities. The wings display notches with occasional

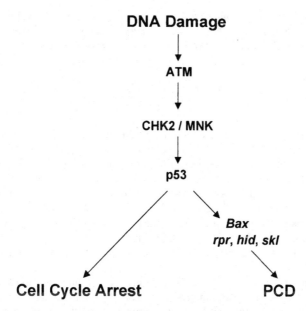

FIGURE 1. A simplified model of DNA damage–induced cellular responses. Core components of the DNA damage signaling in mammals and *Drosophila*. The DNA damage is sensed by ATM. The damage signaling is relayed and amplified by CHK2/MNK kinase. The downstream effector p53 is involved in PCD and cell cycle arrest responses in this signaling cascade. p53 regulates *Bax*, *rpr*, *hid*, and *skl* pro-PCD genes.

blister formation. The defects resulting from the absence of *dATM* suggest it has important developmental functions.

CHK2

Propagation of the DNA damage signal initiated by activation of ATM often involves actions of CHK2, a serine/threonine kinase, which is a primary target of ATM in response to DNA DSBs. For example, CHK2 is not modified or activated in ATM-deficient cells exposed to IR.[19] However, reintroduction of ATM into these cells leads to modification and activation of CHK2 in response to IR.

In humans, mutations in *CHK2* are present in a subset of patients with Li-Fraumeni syndrome who are predisposed to developing diverse types of cancer.[20–22] Li-Fraumeni syndrome is also linked to mutations in *p53*. *CHK2*-deficient mice show no obvious defects under normal conditions.[23] However, thymic tissues of *CHK2*-deficient mice exhibit resistance to IR-induced PCD, while wild-type tissues undergo PCD.[23] Furthermore, after irradiation, *CHK2* mutant mice form tumors earlier than wild-type animals, and they develop a greater number of tumors.[23] Collectively, these findings suggest an important role of CHK2 in DNA damage response signaling.

Drosophila CHK2 (MNK) has a crucial function in the DNA damage response pathway mediating cell cycle arrest and PCD.[24–26] In wild-type wing imaginal discs, IR causes extensive PCD. In contrast, PCD is blocked in *mnk* mutants after irradiation. Reintroduction of an *mnk* wild-type gene restores irradiation-induced PCD. These results demonstrate that MNK is involved in regulating DNA damage–induced PCD. Intriguingly, the MNK protein is expressed in ovaries and germ cell nuclei of the embryo, suggesting that it may act as a safeguard against genetically crippled PGCs.[27]

p53

p53 has been referred to as the guardian of the genome, owing to its roles in various cellular processes necessary to maintain genome integrity and stability.[28] The key function of p53 involves transcriptional regulation. In DNA damage signaling pathways, ATM and CHK2 are primary candidate molecules that aid in activation and stabilization of p53.[29] Downstream targets of p53 are employed in diverse cellular activities, such as *p21* in G1 growth arrest, *GADD45* in DNA repair, and *Bax* in PCD.[30] This suggests important roles for p53 in coordinating cellular responses to DNA damage. Modifications on the p53 protein are likely to regulate its function, although it remains unclear exactly how activation and stabilization of p53 are controlled.[29]

More than 50% of cancerous tumors in humans carry mutations in *p53*.[31] Studies of families with germline *p53* mutations have revealed high incidences of breast cancer, sarcomas of bone and soft tissues, and brain cancer.[32] In mice, loss of *p53* leads to an early onset of spontaneous tumor formation.[33] Involvement of *p53* in DNA damage–induced PCD is demonstrated in the following examples from mice: Intestinal crypt cells undergo PCD in response to IR. However, cells deficient for *p53* are resistant to induction of PCD in response to IR.[34] In culture, *p53* null thymocytes are resistant to IR–induced PCD, while wild-type thymocytes undergo PCD following IR.[35] Finally, wild-type postmitotic CNS neurons, but not *p53* mutant neurons, undergo PCD in response to IR.[36]

Drosophila p53 (*Dmp53*) has also been shown to mediate DNA damage–induced PCD.[37–39] Overexpression of *Dmp53* in the developing eye results in PCD, giving a small eye phenotype. Also, it has been shown that radiation-induced PCD in wing and eye imaginal discs is suppressed by dominant negative *Dmp53* transgene expression.[37,38] Studies of *Dmp53* transcripts during embryogenesis reveal global RNA expression that becomes more restricted to PGCs and hindgut cells during development.[37] Given its central role in DNA damage–induced PCD, it is likely that *Dmp53* plays an important part in regulating germ cell development.

Bax

Bax, a pro-PCD gene belonging to the Bcl-2 family, appears to be an important component in PCD.[40] The Bcl-2 family, defined by the presence of conserved BH domains, includes both pro- and anti-PCD genes. In mammals, interactions of Bcl-2-related proteins are central to death or life decisions of cells. Upon death stimuli, the Bax protein undergoes oligomerization followed by integration into the outer mitochondrial membrane. This relocalization appears important for propagation of death signaling.

Bax is often mutated in human colorectal and hematopoietic malignancies.[41,42] In mice, loss of *Bax* promotes tumor formation.[43] Bax has also been shown to have a role in DNA damage–induced PCD. For example, in *Bax* knockout mice, the dentate gyrus is resistant to IR-induced PCD, whereas IR triggers massive PCD in wild-type animals.[44] *Bax* expression appears to be regulated by p53.[45] Gel retardation assays demonstrate that wild-type but not mutant p53 protein binds to the p53 binding sites of the *Bax* promoter. In *p53*-deficient tumor cells, a wild-type but not a mutant p53 transgene is capable of transactivating the *Bax* promoter to drive the expression of a reporter construct. Furthermore, *Bax* has been shown to be important for *p53*-dependent PCD and contributes to *p53*-dependent tumor suppression in some cells.[46-48]

Though a link between DNA damage–induced PGC PCD and *Bax* has not been described, there is convincing evidence for a role of *Bax* in germ cell PCD.[49,50] Recent findings show that Bax is required for PCD of migratory PGCs in mice. *Bax*-deficient mouse embryos exhibit a substantially increased number of ectopic germ cells compared to wild type.[51] However, the ectopic germ cells later die, indicating the involvement of other PCD pathways.

rpr, skl, and hid

In higher eukaryotes, antagonistic interplay of cell death regulators maintains the fine balance between death and survival of cells. Inhibitors of apoptosis proteins (IAPs) are present to block the proteolytic actions of caspases, key executers of PCD. In *Drosophila*, *rpr*, *grim*, *hid* and *skl* are IAP antagonists that play major roles in promoting PCD.[52] Recently, it has been proposed that *rpr*, *hid*, and *skl* function in IR-induced PCD on the basis of the observation that flies deficient for *rpr*, *skl*, and *hid* exhibit reduced IR-induced PCD in wing imaginal discs.[39]

rpr has been shown to be involved in Dmp53-mediated responses to radiation.[39,55] Dmp53 binds to a radiation-responsive *cis*-regulatory region of *rpr*. After IR, the *rpr* radiation-responsive element is activated in *Dmp53* wild-type embryos, but not in *Dmp53*-deficient embryos.[38] Genome-wide analysis of IR-induced transcription shows that *rpr* is induced following IR.[39] However, this is not the case in *Dmp53* or *mnk* (*CHK2*) mutant backgrounds.

Similarly, microarray profiles of IR-responsive genes reveal that *skl* and *hid* are highly induced by IR in wild-type, but not *Dmp53* or *mnk* (*CHK2*)–deficient backgrounds.[39] Therefore, *skl* and *hid* are potential targets of MNK(CHK2)/Dmp53-mediated damage signaling.

CONCLUDING REMARKS

The previous sections present evidence that *ATM*, *CHK2/mnk*, *p53*, *Bax*, *rpr*, *hid* and *skl* all play important roles in DNA damage–induced PCD. Are these core components of DNA damage signaling a part of PGC development helping to ensure elimination of genetically damaged germ cells? ATM expression has not been reported in PGCs. Involvement of *CHK2* and *p53* in mammalian PGC development are yet to be analyzed. However, MNK/CHK2 protein and *Dmp53* mRNA are expressed in *Drosophila* PGCs during embryogenesis, making it likely that they have functions in this particular developmental context. Bax has been shown to be involved in PGC PCD in mice. Whether Bax functions in DNA damage–induced PCD of PGCs to select against cells carrying defective DNA remains an open question. Intriguingly, there is evidence indicating *CHK2*- and *p53*-dependent activation of *Bax* in response to IR.[54, 55] So far, no published studies show involvement of *rpr*, *hid* and *skl* in germ cell development. However, *rpr*, *hid* and *skl* transcript levels increase in *mnk*- and *Dmp53*-dependent manners following IR. Future studies will determine involvement of these pro-PCD molecules and their interactions with other components of the networks in PGC PCD in developing embryos.

REFERENCES

1. RODRIGUEZ, I. *et al.* 1997. An early and massive wave of germinal cell apoptosis is required for the development of functional spermatogenesis. EMBO J. **16:** 2262–2270.
2. SCHNEIDER, D.T. *et al.* 2001. Multipoint imprinting analysis indicates a common precursor cell for gonadal and nongonadal pediatric germ cell tumors. Cancer Res. **61:** 7268–7276.
3. FURUCHI, T. *et al.* 1996. Inhibition of testicular germ cell apoptosis and differentiation in mice misexpressing Bcl-2 in spermatogonia. Development **122:** 1703–1709.
4. ZHOU, B.B. & S.J. ELLEDGE. 2000. The DNA damage response: putting checkpoints in perspective. Nature **408:** 433–439.
5. OLIVE, P.L. 1998. The role of DNA single- and double-strand breaks in cell killing by ionizing radiation. Radiat Res. **150:** S42–51.
6. HOEIJMAKERS, J.H. 2001. Genome maintenance mechanisms for preventing cancer. Nature **411:** 366–374.
7. VAN GENT, D., J.H. HOEIJMAKERS & R. KANAAR. 2001. Chromosomal stability and the DNA double-stranded break connection. Nat. Rev. Genet. **2:** 196–206.
8. ELLEDGE, S.J. 1996. Cell cycle checkpoints: preventing an identity crisis. Science **274:** 1664–1672.
9. VALERIE, K. & L.F. POVIRK. 2003. Regulation and mechanisms of mammalian double-strand break repair. Oncogene **22:** 5792–5812.
10. SHILOH, Y. 2003. ATM and related protein kinases: safeguarding genome integrity. Nat. Rev. Cancer. **3:** 155–168.
11. BAKKENIST, C.J. & M.B. KASTAN. 2003. DNA damage activates ATM through intermolecular autophosphorylation and dimer dissociation. Nature **421:** 499–506.
12. ANDEGEKO, Y. *et al.* 2001. Nuclear retention of ATM at sites of DNA double strand breaks. J. Biol. Chem. **276:** 38224–38230.
13. HARNDEN, D.G. 1994. The nature of ataxia-telangiectasia: problems and perspectives. Int. J. Radiat. Biol. **66:** S13–19.
14. SHILOH, Y. 1995. Ataxia-telangiectasia: closer to unraveling the mystery. Eur. J. Hum. Genet. **3:** 116–138.
15. KULJIS, R.O. *et al.* 1997. Degeneration of neurons, synapses, and neuropil and glial activation in a murine Atm knockout model of ataxia-telangiectasia. Proc. Natl. Acad. Sci. USA. **94:** 12688–12693.
16. XU, Y. *et al.* 1996. Targeted disruption of ATM leads to growth retardation, chromosomal fragmentation during meiosis, immune defects, and thymic lymphoma. Genes Dev. **10:** 2411–2422.
17. HERZOG, K.H. *et al.* 1998. Requirement for Atm in ionizing radiation-induced cell death in the developing central nervous system. Science **280:** 1089–1091.
18. SONG, Y.H. *et al.* 2004. The *Drosophila* ATM ortholog, dATM, mediates the response to ionizing radiation and to spontaneous DNA damage during development. Curr. Biol. **14:** 1354–1359.
19. MATSUOKA, S., M. HUANG & S.J. ELLEDGE. 1998. Linkage of ATM to cell cycle regulation by the Chk2 protein kinase. Science **282:** 1893–1897.
20. BELL, D.W. *et al.* 1999. Heterozygous germ line hCHK2 mutations in Li-Fraumeni syndrome. Science **286:** 2528–2531.

21. LEE, S.B. *et al.* 2001. Destabilization of CHK2 by a missense mutation associated with Li-Fraumeni syndrome. Cancer Res. **61:** 8062–8067.
22. VAHTERISTO, P. *et al.* 2001. p53, CHK2, and CHK1 genes in Finnish families with Li-Fraumeni syndrome: further evidence of CHK2 in inherited cancer predisposition. Cancer Res. **61:** 5718–5722.
23. HIRAO, A. *et al.* 2002. Chk2 is a tumor suppressor that regulates apoptosis in both an ataxia telangiectasia mutated (ATM)-dependent and an ATM-independent manner. Mol. Cell. Biol. **22:** 6521–6532.
24. XU, J., S. XIN & W. DU. 2001. *Drosophila* Chk2 is required for DNA damage-mediated cell cycle arrest and apoptosis. FEBS Lett. **508:** 394–398.
25. PETERS, M. *et al.* 2002. Chk2 regulates irradiation-induced, p53-mediated apoptosis in *Drosophila*. Proc. Natl. Acad. Sci. USA **99:** 11305–11310.
26. MASROUHA, N. *et al.* 2003. The *Drosophila* chk2 gene loki is essential for embryonic DNA double-strand-break checkpoints induced in S phase or G2. Genetics **163:** 973–982.
27. OISHI, I. *et al.* 1998. A novel *Drosophila* nuclear protein serine/threonine kinase expressed in the germline during its establishment. Mech. Dev. **71:** 49–63.
28. LANE, D.P. 1992. Cancer. p53, guardian of the genome. Nature. **358:** 15–16.
29. SUTCLIFFE, J. E. & A. BREHM. 2004. Of flies and men; p53, a tumour suppressor. FEBS Lett. **567:** 86–91.
30. VOUSDEN, K.H. & X. LU. 2002. Live or let die: the cell's response to p53. Nat. Rev. Cancer **2:** 594–604.
31. GREENBLATT, M.S. *et al.* 1994. Mutations in the p53 tumor suppressor gene: clues to cancer etiology and molecular pathogenesis. Cancer Res. **54:** 4855–4878.
32. KLEIHUES, P. *et al.* 1997. Tumors associated with p53 germline mutations: a synopsis of 91 families. Am. J. Pathol. **150:** 1–13.
33. GHEBRANIOUS, N. & L.A. DONEHOWER. 1998. Mouse models in tumor suppression. Oncogene **17:** 3385–3400.
34. CLARKE, A.R. *et al.* 1994. p53 dependence of early apoptotic and proliferative responses within the mouse intestinal epithelium following gamma-irradiation. Oncogene **9:** 1767–1773.
35. CLARKE, A.R. *et al.* 1993. Thymocyte apoptosis induced by p53-dependent and independent pathways. Nature **362:** 849–852.
36. ENOKIDO, Y. *et al.* 1996. Involvement of p53 in DNA strand break-induced apoptosis in postmitotic CNS neurons. Eur. J. Neurosci. **8:** 1812–1821.
37. OLLMANN, M. *et al.* 2000. *Drosophila* p53 is a structural and functional homolog of the tumor suppressor p53. Cell **101:** 91–101.
38. BRODSKY, M.H. *et al.* 2000. *Drosophila* p53 binds a damage response element at the reaper locus. Cell **101:** 103–113.
39. BRODSKY, M.H. *et al.* 2004. *Drosophila* melanogaster MNK/Chk2 and p53 regulate multiple DNA repair and apoptotic pathways following DNA damage. Mol. Cell. Biol. **24:** 1219–1231.
40. KIRKIN, V., S. JOOS & M. ZORNIG. 2004. The role of Bcl-2 family members in tumorigenesis. Biochim. Biophys. Acta **1644:** 229–249.
41. RAMPINO, N. *et al.* 1997. Somatic frameshift mutations in the BAX gene in colon cancers of the microsatellite mutator phenotype. Science **275:** 967–969.
42. MEIJERINK, J.P. *et al.* 1998. Hematopoietic malignancies demonstrate loss-of-function mutations of BAX. Blood **91:** 2991–2997.
43. IONOV, Y. *et al.* 2000. Mutational inactivation of the proapoptotic gene BAX confers selective advantage during tumor clonal evolution. Proc. Natl. Acad. Sci. USA **97:** 10872–10877.
44. CHONG, M.J. *et al.* 2000. Atm and Bax cooperate in ionizing radiation-induced apoptosis in the central nervous system. Proc. Natl. Acad. Sci. USA **97:** 889–894.
45. MIYASHITA, T. & J.C. REED. 1995. Tumor suppressor p53 is a direct transcriptional activator of the human bax gene. Cell **80:** 293–299.
46. YIN, C. *et al.* 1997. Bax suppresses tumorigenesis and stimulates apoptosis in vivo. Nature **385:** 637–640.
47. CREGAN, S.P. *et al.* 1999. Bax-dependent caspase-3 activation is a key determinant in p53-induced apoptosis in neurons. J. Neurosci. **19:** 7860–7869.

48. XIANG, H. *et al.* 1998. Bax involvement in p53-mediated neuronal cell death. J. Neurosci. **18:** 1363–1373.
49. PEREZ, G.I. *et al.* 1999. Prolongation of ovarian lifespan into advanced chronological age by Bax-deficiency. Nat. Genet. **21:** 200–203.
50. RUCKER, E.B., 3RD, *et al.* 2000. Bcl-x and Bax regulate mouse primordial germ cell survival and apoptosis during embryogenesis. Mol. Endocrinol. **14:** 1038–1052.
51. STALLOCK, J. *et al.* 2003. The pro-apoptotic gene Bax is required for the death of ectopic primordial germ cells during their migration in the mouse embryo. Development **130:** 6589–6597.
52. BERGMANN, A., A.Y. YANG & M. SRIVASTAVA. 2003. Regulators of IAP function: coming to grips with the grim reaper. Curr. Opin. Cell Biol. **15:** 717–724.
53. SOGAME, N., M. KIM & J.M. ABRAMS. 2003. *Drosophila* p53 preserves genomic stability by regulating cell death. Proc. Natl. Acad. Sci. USA **100:** 4696–4701.
54. TAKAI, H. *et al.* 2002. Chk2-deficient mice exhibit radioresistance and defective p53-mediated transcription. EMBO J. **21:** 5195–5205.
55. HIRAO, A. *et al.* 2000. DNA damage-induced activation of p53 by the checkpoint kinase Chk2. Science **287:** 1824–1827.

G Protein–Coupled Receptor Roles in Cell Migration and Cell Death Decisions

ANGELA R. KAMPS AND CLARK R. COFFMAN

Department of Genetics, Development and Cell Biology, Iowa State University, Ames, Iowa 50011, USA

ABSTRACT: Recognition of external conditions and the elicitation of appropriate responses are critical to a cell's ability to adjust to various developmental and environmental cues. G protein–coupled receptors (GPCRs) are a large class of receptors that act to relay external information into the cell by initiating signaling pathways that allow the cell to adapt to its present conditions. There are numerous ligands that activate GPCRs to initiate a multitude of intracellular signaling cascades involved in critical decisions including cell growth, differentiation, proliferation, migration, survival, and death. This article focuses on the signaling pathways involved in cell migration, survival, and death decisions with an emphasis on germ cells from various organisms.

KEYWORDS: G protein–coupled receptors; CXCR4; SDF-1; programmed cell death; apoptosis; cell migration; development; *Drosophila melanogaster*

INTRODUCTION

Migrating cells contact and interact with numerous cell types and substrates. The ability of these cells to react to an ever-changing environment is dependent on the recognition of external stimuli, followed by the activation of the appropriate responses. G protein–coupled receptors (GPCRs) are one of the mechanisms through which environmental cues are relayed into the cell in order to elicit a reaction. GPCRs are an extremely large class of proteins found in essentially all cell types. GPCRs recognize and respond to a wide variety of external signals including organic molecules, ions, proteases, neurotransmitters, peptides, glycoproteins, hormones, nucleotides, lipids, and light.[1–3] The ligand can interact with extracellular regions of the GPCR, the transmembrane domains, or both. A conformational change in the seven-transmembrane helical structure results in the activation of the GPCR, allowing it to interact with the appropriate heterotrimeric G protein within the cell. GPCRs relay signals into the cell, utilizing a plethora of intracellular signal transduction pathways, and initiate signaling cascades that mediate diverse functions including cell growth, differentiation, proliferation, migration, survival, and death.[4,5] In addi-

Address for correspondence: Angela R. Kamps, Department of Genetics, Development and Cell Biology, Iowa State University, 3238 Molecular Biology Building, Ames, Iowa 50011-3260. Voice: 515-294-4734; fax: 515-294-6755.
amortved@iastate.edu

Ann. N.Y. Acad. Sci. 1049: 17–23 (2005). © 2005 New York Academy of Sciences.
doi: 10.1196/annals.1334.003

tion, kinases, proteases, adaptor proteins, and ion channels often act within these pathways to modulate the actions of GPCRs.[1-3]

This review will focus on GPCR-mediated cell migration and programmed cell death, emphasizing their roles in germ cell development. Current models of germ cell migration in mice, zebrafish, chick, and *Drosophila* will be presented as will the recently identified roles of GPCRs in the action of lysophospholipids in cell survival pathways.[5] In addition, recent information will be discussed on the *Drosophila* GPCR, Tre1, which has been shown to have a role in germ cell migration and programmed cell death during *Drosophila* development.[6,7]

GPCRs IN CELL MIGRATION

GPCRs mediate the successful migration of numerous cell types in development, wound healing, and immune responses.[8-10] In early development, the germline stem cells must navigate through multiple cell types and substrates to reach their ultimate destinations, the gonads.[8,11,12] Population of the gonads with germ cells is essential to establish the next generation. Recently, a conserved molecular mechanism involving the CXCR4/SDF-1 receptor/ligand pair has been identified in mouse, chick, and zebrafish germ cell development.[13-18] The chemoattractant stromal cell–derived factor 1 (SDF-1) and its GPCR counterpart, CXCR4, are known to function in many other developmental, homeostatic, and immune system response processes. Elucidation of the mechanisms of these signaling pathways will be a fruitful area of research in the near future as connections to other cell processes will no doubt be uncovered.

Murine germ cells originate during gastrulation.[19] Time-lapse analysis studies using immunofluorescence of the germ cells have revealed many details of murine germ cell migration.[20,21] The migratory phase of these cells begins shortly after their formation as they move through the primitive streak on their way to the hindgut. While contained within the hindgut, the germ cells are highly motile, but no directed movement is observed until they cross the hindgut epithelium and begin to move toward the genital ridges. The germ cells compact into a cluster as migration nears completion and interact and coalesce with somatic gonadal precursor cells to form the gonad.[21]

Recently, it has been reported that the CXCR4/SDF-1 ligand/receptor pair is involved in the final stages of murine germ cell migration, as the germ cells traverse the hindgut epithelium and migrate towards the genital ridges.[13] SDF-1 appears to have no effect on germ cell movements prior to crossing the gut epithelium. The GPCR CXCR4 is expressed in the migrating germ cells and is able to respond to SDF-1 being expressed by the dorsal body wall and the genital ridges. The hypothesis that CXCR4 and SDF-1 are part of a germ cell chemoattractant mechanism in mice has been further supported by misexpression and knockdown experiments. When mouse mutants in either the CXCR4 receptor or the SDF-1 ligand are generated, successful incorporation of germ cells into the gonads is severely impaired.[13,14] Movement of germ cells lacking the CXCR4 receptor appears normal up to their incorporation into the hindgut. In the ensuing migratory steps, however, the germ cells are still observed on the migration path and few have reached the genital ridges at a time when wild-type germ cells have completed migration. In addition, misexpression of SDF-1 can attract some germ cells to ectopic locations and impede the directed migration of germ cells towards the endogenous ligand.[13]

Germ cell migration in chick embryos is comparable to leukocyte migration in that germ cells utilize the vascular system for the initial passive migration steps on their way to somatic gonad tissue.[15] The germ cells bind to the vasculature adjacent to the somatic gonad precursor cells. They must then pass through the blood vessel endothelium in order to migrate to their target. As in mouse germ cell migration, SDF-1 appears to be a ligand that provides an attractive signal for the germ cells after crossing this epithelial layer. SDF-1 mRNA is expressed along the post-endothelial migration path of germ cells. Ectopic expression of SDF-1α causes aberrant germ cell migration and the accumulation of germ cells around the site of SDF-1α expression.[15]

In zebrafish germ cells, SDF-1a appears to be required throughout germ cell migration.[16–18] First, SDF-1a and the germ cell-expressed receptor, CXCR4b, are necessary for directed germ cell movements during the initial migrations of germ cells from random origination positions in the embryo. Second, this receptor–ligand pair aids in movement to, and coalescence with, the somatic gonad cells in the final stages of migration.[11] The somatic cells that act as intermediate targets of the germ cells express SDF-1a. Both loss-of-function and ectopic expression experiments support the conclusion that SDF-1a acts as an attractive signal for directed germ cell migration. Mutations that disrupt SDF-1a expression by altering somatic cell patterning affect the successful migration of zebrafish germ cells.[16] Inhibiting translation of SDF-1a protein using morpholinos also disrupts germ cell migration. In these mutants, zebrafish germ cells are unable to consistently locate the gonads and scatter to ectopic locations.[16] In additional experiments, it was found that ectopic SDF-1a was able to attract CXCR4b-expressing germ cells when endogenous SDF1a levels were reduced.[16] In loss-of-function experiments where the gene encoding the CXCR4b receptor was mutated, germ cells scattered to ectopic locations rather than clustering at the gonadal anlagen.[18] Recent research in zebrafish has begun to elucidate the downstream components of the SDF-1a/CXCR4b interaction. Inhibition of the G protein, G_i, results in a phenotype similar to loss of function *sdf-1a* or *cxcr4b* mutants.[17]

GPCRs IN REGULATION OF CELL DEATH AND CELL SURVIVAL DECISIONS

Another crucial role for GPCRs in development is in the regulation of cell survival and cell death. The extensive networks of cellular signaling pathways connected to GPCRs allow fine control of cell survival and death in a context-dependent manner. Recently, two lysophospholipids—sphingosine 1-phosphate (S1P) and lysophosphatidic acid (LPA)—have been identified as GPCR ligands.[5] The ligands signal through their respective receptor subtypes and the cellular response depends upon the cell type and/or the cellular context. In mammals, five S1P receptors ($S1P_{1-5}$) and four LPA receptors (LPA_{1-4}) have been identified.[4,5] The S1P receptors are extremely selective in their ligand choice as they only recognize S1P and dihydroS1P.[22] Both the LPA and S1P receptor subtypes have the ability to bind different types of G protein alpha subunits including $G_{\alpha i/o}$, $G_{\alpha q/11}$, and $G_{\alpha 12/13}$.[4,5] The difference in temporal and spatial expression patterns of the G_α subunits allows for control of multiple activities. However, regulation of cell survival appears to most often be regulated through the $G_{\alpha i}$ protein.

Apart from their role in maintaining membrane structure, S1P and other sphingolipids are key players in cell growth, cell survival, and cell death.[22,23] S1P can act either as a ligand for S1P receptors or as a second messenger within the cell.[5,22] Extracellular S1P function is mediated through its GPCR and has been shown to promote cell survival in melanocytes, neutrophils and leukemia cells.[5] Interestingly, S1P can also protect cells without acting through its GPCR. Overexpression of S1P or sphingosine kinase, the enzyme that converts sphingosine into S1P, suppresses cell death, even in the absence of ligand-activated S1P function.[24–26] For example, in mouse embryonic fibroblast cells that have lost S1P-receptor function, increased levels of S1P due to overexpression of sphingosine kinase protect these cells from death.[26]

In contrast, the S1P precursors ceramide and sphingosine are pro-apoptotic. It has been suggested that the balance between intracellular S1P and its precursors determines the fate of some cells.[27] Higher concentrations of S1P favor cell survival, while an abundance of ceramide and sphingosine leads to termination of the cell.

LPA has opposing roles in cell survival decisions depending on the cell type. LPA is anti-apoptotic in Schwann cells and ovarian cancer cells.[4,28] Loss of LPA1 receptor function in these cells results in increased levels of apoptosis, indicating a requirement for LPA in the survival of these cells. However, in TF-1 hematopoietic cells or hippocampal neurons, LPA is pro-apoptotic.[29,30] In some T lymphoblasts, LPA can elicit both apoptotic and pro-survival responses within the same cell type depending on the presence or absence of other regulatory molecules.[4]

GPCRs IN *DROSOPHILA* GERMLINE DEVELOPMENT

Drosophila germ cell development also requires GPCR function.[6] Similar to mouse and chick germ cell development, *Drosophila* germ cells traverse an epithelial layer, the posterior midgut epithelium.[31–34] This is followed by migration through mesodermal cell layers and coalescence with somatic gonadal precursor cells to form the gonads (FIG. 1A). *Drosophila* germ cells require a GPCR encoded by the *tre1* gene. Maternally expressed *tre1* has several roles that include initiating the crossing of the midgut epithelium, pathfinding to the somatic gonad cells, and regulation of programmed cell death.[7,35] In embryos lacking both maternal and zygotic *tre1* expression, the germ cells remain trapped within the primordial midgut.[6] Zygotic expression of *tre1* can partially rescue the germ cell migration phenotype of embryos from *tre1* mutant mothers.[6] Unlike mouse, chick, and zebrafish germ cells, where SDF-1 has been convincingly identified as a ligand of CXCR4,[15,16,18,36–39] the ligand for the GPCR located on the germ cell membrane of *Drosophila* remains elusive.

The *scattershot* (*sctt*) mutation, an ethylmethane sulfonate (EMS)-induced partial loss-of-function allele of *tre1,* reveals roles for *tre1* in both germ cell migration and programmed cell death.[7] In *sctt* mutants, the germ cells initiate migration by crossing the posterior midgut epithelium. However, directed migration to the gonads is disrupted and the germ cells scatter throughout the posterior half of the embryo (FIG.

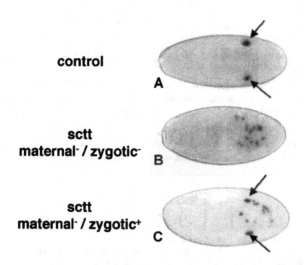

control

A

sctt
maternal⁻ / zygotic⁻

B

sctt
maternal⁻ / zygotic⁺

C

FIGURE 1. The *scattershot* (*sctt*) phenotypes. The *sctt* embryos have defects in both germ cell migration and programmed cell death. In all panels, dorsal views of 12–15-hour embryos are shown. Anterior is to the *left*. Germ cells are labeled using a *fat facets-lacZ* transgene. (**A**) Control embryo. At 12–15 hours of development, the germ cells have reached the somatic gonad precursor cells and have coalesced to form the gonads. (**B**) An embryo from a homozygous mutant mother that also lacks zygotic function of *sctt*. In these mutants, the germ cells lack directed migration and scatter throughout the posterior half of the embryo. In addition, the germ cells ectopic to the gonads fail to undergo programmed cell death. The somatic gonad forms normally. (**C**) An embryo from a homozygous mutant mother that has been crossed to a male harboring a wild-type copy of the *tre1* gene. The germ cell migration defect is completely rescued. However, germ cells ectopic to the gonads remain.

1B). Embryos from *sctt* mutant mothers also display a programmed cell death defect. During development of wild-type *Drosophila melanogaster,* approximately 50% of the germ cells that originate at the posterior pole fail to reach the gonads and are eradicated during migration.[32,33] While the signaling events that promote cell death and/or survival and the exact timing of programmed cell death in wild-type *Drosophila* embryos are unknown, few germ cells remain ectopic to the gonads in wild-type embryos after the germ cells and the somatic gonad cells have coalesced (FIG. 1A). The germ cell migration and programmed cell death defects of *sctt* mutants can be genetically uncoupled by zygotic rescue. When a wild-type copy of *tre1* is supplied paternally, germ cell migration is restored but germ cells ectopic to the gonads still fail to complete the cell death program (FIG. 1C). Elucidation of the upstream and downstream components of the *tre1*-mediated signaling pathway will be crucial to our understanding of germ cell migration in *Drosophila*.

The multitudes of cellular processes mediated through the GPCR superfamily suggest an amazingly complex and well-regulated mechanism for the transduction of stimuli into various independent yet entangled cellular responses. The role of *tre1* in *Drosophila* germ cell development is interesting as it demonstrates that multiple developmental functions are regulated through a single GPCR. One could imagine the usefulness of integrated pathways for cell death and cell migration. It would be beneficial for elimination of cells that are in excess, are misplaced, or fall behind during the pathfinding processes critical to so many developmental stages. The inability to efficiently rid the organism of ectopic cells is often detrimental. In humans, ectopic germ cells are etiologic in a variety of tumors including teratomas, endodermal sinus tumors, embryonal carcinomas, choriocarcinomas, and testicular and ovarian carcinomas.[40–43] Further research on GPCR-mediated signaling in various model organisms will undoubtedly deepen our understanding of the regulatory mechanisms controlling cell migration and programmed cell death and the common machinery that allows for crosstalk between these important cellular functions.

REFERENCES

1. GETHER, U. 2000. Uncovering molecular mechanisms involved in activation of G protein-coupled receptors. Endocr. Rev. **21:** 90-113.
2. KARNIK, S.S. *et al.* 2003. Activation of G-protein-coupled receptors: a common molecular mechanism. Trends Endocrinol. Metab. **14:** 431-437.
3. RADER, A.J. *et al.* 2004. Identification of core amino acids stabilizing rhodopsin. Proc. Natl. Acad. Sci. USA **101:** 7246-7251.
4. YE, X. *et al.* 2002. Lysophosphatidic acid as a novel cell survival/apoptotic factor. Biochim. Biophys. Acta **1585:** 108-113.
5. RADEFF-HUANG, J. *et al.* 2004. G protein mediated signaling pathways in lysophospholipid induced cell proliferation and survival. J. Cell. Biochem. **92:** 949-966.
6. KUNWAR, P.S. *et al.* 2003. Tre1, a G protein-coupled receptor, directs transepithelial migration of *Drosophila* germ cells. PLoS Biol. **1:** E80.
7. COFFMAN, C.R. *et al.* 2002. Identification of X-linked genes required for migration and programmed cell death of *Drosophila melanogaster* germ cells. Genetics **162:** 273-284.
8. SANTOS, A.C. & R. LEHMANN. 2004. Germ cell specification and migration in *Drosophila* and beyond. Curr. Biol. **14:** R578-89.
9. SMITH, J.M. *et al.* 2005. CXCL12 activation of CXCR4 regulates mucosal host defense through stimulation of epithelial cell migration and promotion of intestinal barrier integrity. Am. J. Physiol. Gastrointest. Liver Physiol. **288:** G316–326 [Epub 2004 Sept. 9. PMID: 15358596].
10. SPRINGER, T.A. 1995. Traffic signals on endothelium for lymphocyte recirculation and leukocyte emigration. Annu. Rev. Physiol. **57:** 827-872.
11. RAZ, E. 2004. Guidance of primordial germ cell migration. Curr. Opin. Cell Biol. **16:** 169-173.
12. MOLYNEAUX, K. & C. WYLIE. 2004. Primordial germ cell migration. Int. J. Dev. Biol. **48:** 537-543.
13. MOLYNEAUX, K.A. *et al.* 2003. The chemokine SDF1/CXCL12 and its receptor CXCR4 regulate mouse germ cell migration and survival. Development. **130:** 4279-4286.
14. ARA, T. *et al.* 2003. Impaired colonization of the gonads by primordial germ cells in mice lacking a chemokine, stromal cell-derived factor-1 (SDF-1). Proc. Natl. Acad. Sci. USA **100:** 5319-5323.
15. STEBLER, J. *et al.* 2004. Primordial germ cell migration in the chick and mouse embryo: the role of the chemokine SDF-1/CXCL12. Dev. Biol. **272:** 351-361.
16. DOITSIDOU, M. *et al.* 2002. Guidance of primordial germ cell migration by the chemokine SDF-1. Cell **111:** 647-659.
17. DUMSTREI, K., R. MENNECKE & E. RAZ. 2004. Signaling pathways controlling primordial germ cell migration in zebrafish. J. Cell Sci. **117:** 4787-4795.
18. KNAUT, H. *et al.* 2003. A zebrafish homologue of the chemokine receptor Cxcr4 is a germ-cell guidance receptor. Nature **421:** 279-282.
19. GINSBURG, M., M.H. SNOW & A. MCLAREN. 1990. Primordial germ cells in the mouse embryo during gastrulation. Development **110:** 521-528.

20. ANDERSON, R., et al. 2000. The onset of germ cell migration in the mouse embryo. Mech Dev. **91:** 61-68.
21. MOLYNEAUX, K.A., et al. 2001. Time-lapse analysis of living mouse germ cell migration. Dev. Biol. **240:** 488-498.
22. SPIEGEL, S. & R. KOLESNICK. 2002. Sphingosine 1-phosphate as a therapeutic agent. Leukemia. **16:** 1596-1602.
23. KOLESNICK, R.N. 1987. 1,2-Diacylglycerols but not phorbol esters stimulate sphingomyelin hydrolysis in GH3 pituitary cells. J. Biol. Chem. **262:** 16759-16762.
24. VAN BROCKLYN, J.R. et al. 1998. Dual actions of sphingosine-1-phosphate: extracellular through the Gi-coupled receptor Edg-1 and intracellular to regulate proliferation and survival. J. Cell Biol. **142:** 229-240.
25. EDSALL, L.C., et al. 2001. Sphingosine kinase expression regulates apoptosis and caspase activation in PC12 cells. J. Neurochem. **76:** 1573-1584.
26. OLIVERA, A., et al. 2003. Sphingosine kinase type 1 induces G12/13-mediated stress fiber formation, yet promotes growth and survival independent of G protein-coupled receptors. J. Biol. Chem. **278:** 46452-46460.
27. SPIEGEL, S. & S. MILSTIEN. 2003. Sphingosine-1-phosphate: an enigmatic signalling lipid. Nat. Rev. Mol. Cell. Biol. **4:** 397-407.
28. CONTOS, J. J., et al. 2000. Requirement for the lpA1 lysophosphatidic acid receptor gene in normal suckling behavior. Proc. Natl. Acad. Sci. USA **97:** 13384-13389.
29. LAI, J. M., C. L. HSIEH & Z. F. CHANG. 2003. Caspase activation during phorbol ester-induced apoptosis requires ROCK-ependent myosin-mediated contraction. J. Cell Sci. **116:** 3491-3501.
30. HOLTSBERG, F. W. et al. 1998. Lysophosphatidic acid induces necrosis and apoptosis in hippocampal neurons. J. Neurochem. **70:** 66-76.
31. CAMPOS-ORTEGA, J.V.H. 1997. The Embryonic Development of *Drosophila melanogaster*. Springer-Verlag. New York/Berlin.
32. SONNENBLICK, B.P. 1941. Germ cell movements and sex differentiation of the gonads in the *Drosophila* embryo. Proc. Natl. Acad. Sci. USA **27:** 484-489.
33. SONNENBLICK, B.P. 1950. The Early Embryology of *Drosophila melanogaster*. Wiley. New York.
34. Moore, L.A. et al. 1998. Identification of genes controlling germ cell migration and embryonic gonad formation in *Drosophila*. Development **125:** 667-678.
35. KUNWAR, P.S. & R. LEHMANN. 2003. Developmental biology: germ-cell attraction. Nature **421:** 226-227.
36. BAGGIOLINI, M., B. DEWALD & B. MOSER. 1997. Human chemokines: an update. Annu. Rev. Immunol. **15:** 675-705.
37. BLEUL, C.C. et al. 1996. The lymphocyte chemoattractant SDF-1 is a ligand for LESTR/fusin and blocks HIV-1 entry. Nature **382:** 829-833.
38. BLEUL, C.C. et al. 1996. A highly efficacious lymphocyte chemoattractant, stromal cell-derived factor 1 (SDF-1). J. Exp. Med. **184:** 1101-1109.
39. MA, Q. et al. 1998. Impaired B-lymphopoiesis, myelopoiesis, and derailed cerebellar neuron migration in CXCR4- and SDF-1-deficient mice. Proc. Natl. Acad. Sci. USA **95:** 9448-9453.
40. FAUCI, A.S., E. BRAUNWALD, K.J. ISSELBACHER, et al. 1998. Harrison's Principles of Internal Medicine. McGraw-Hill. New York.
41. GATCOMBE, H.G, et al. 2004. Primary retroperitoneal teratomas: a review of the literature. J. Surg. Oncol. **86:** 107-113.
42. ULBRIGHT, T.M. 2004. Gonadal teratomas: a review and speculation. Adv. Anat. Pathol. **11:** 10-23.
43. BRANDES, A.A., L.M. PASETTO & S. MONFARDINI. 2000. The treatment of cranial germ cell tumours. Cancer Treat. Rev. **26:** 233-242.

Growth and Differentiation of Astrocytes and Neural Progenitor Cells on Micropatterned Polymer Films

JENNIFER B. RECKNOR,[a,c] D. S. SAKAGUCHI,[b,c] AND
S. K. MALLAPRAGADA[a,c]

Department of [a]Chemical Engineering, Department of [b]Genetics, Development, and Cell Biology, and [c]The Neuroscience Program, Iowa State University, Ames, Iowa 50011,USA

ABSTRACT: This paper investigates the influence of micropatterned polymers and chemical modification on neural progenitor cell growth and differentiation in co-culture systems with astrocytes. We sought to develop stategies to facilitate nerve regeneration using a synergistic combination of guidance cues, investigating the cellular mechanisms of nerve repair using adult rat hippocampal progenitor cells (AHPCs). Our studies have shown that this synergistic combination of physical, chemical, and biological cues can lead to oriented growth of astrocytes and progenitor cells, can control and accelerate neurite outgrowth and alignment *in vitro*, and may influence differentiation of progenitor cells.

KEYWORDS: Schwann cell; neural progenitor cell; micropatterning; photolithography; laminin; nerve regeneration

A key factor in central nervous system (CNS) regeneration is directed neuronal outgrowth. *In vivo*, extracellular matrix and oriented tissue structures influence cell migration and axon outgrowth. *In vitro*, cells recognize three-dimensional geometric configurations of substrate surfaces and their growth can be controlled and guided through the fabrication of microgrooves and other patterns on these surfaces. Cellular response to the surrounding topography depends on the characteristics of the substrate, including the types, densities, and magnitudes of the features, as well as cell type and interactions with neighboring cells.

Combining physical, chemical, and biological guidance cues potentially generates a supportive environment for eliciting regeneration that may lead to the restoration of neural function in the injured or diseased CNS. Neural progenitor cells have been shown to differentiate *in vitro* and express a variety of neuronal proteins and morphologies. In an effort to develop a regenerative environment following injury to the CNS, we are investigating directional growth, differentiation, and expansion *in vitro* of enhanced green fluorescent protein (eGFP)–expresssing adult rat hippocam-

Address for correspondence: S.K. Mallapragada, 3035 Sweeney Hall, Iowa State University, Ames, IA 50011. Voice: + 1-515-294-7407; fax + 1-515-294-2689.
suryakm@iastate.edu

Ann. N.Y. Acad. Sci. 1049: 24–27 (2005). © 2005 New York Academy of Sciences.
doi: 10.1196/annals.1334.004

pal progenitor cells (AHPCs developed and donated by F. Gage, The Salk Institute) on micropatterned polymer substrates chemically modified with adhesive proteins.

Manipulating a combination of guidance cues, we are working to direct AHPC differentiation and expansion as well as to control and accelerate AHPC outgrowth and alignment *in vitro*. To provide physical guidance, micropatterned polymer substrates of polystyrene (PS) were fabricated by means of conventional photolithographic techniques and reactive ion etching. Thin PS films (~50-70 μm) were created by solvent casting a 7–10% (w/v) PS (MW 125,000–250,000) solution in toluene onto a photolithographically etched silicon wafer having the desired micropatterns.[1] The patterns used for these polymer films are described by: groove width (μm)/groove spacing (or mesa width) (μm)/groove depth (μm). The micropattern dimensions used for these experiments were 10/20/3 μm (PS) for astrocytes and 16/13/4 μm (PS) (FIG. 1A.) for AHPCs and astrocyte-AHPC co-cultures. Poly-L-lysine (PLL) (0.1 mg/mL in borate buffer) and laminin [0.01 mg/mL in Earle's balanced salt solution (EBSS)] were selectively adsorbed to the microgrooves using a surface tension–based technique.[2]

In an effort to develop a permissive environment that may influence neural progenitor cell differentiation, directional growth of astrocytes was achieved on polymer substrates *in vitro*. A population of purified postnatal rat type-1 astrocytes was obtained from the cerebral cortex of neonatal rat pups as described in Recknor *et al.*[1] Cells were purified and maintained in modified minimum essential medium (MMEM), consisting of minimum essential medium (MEM) supplemented with 10% v/v fetal bovine serum (FBS). Astrocytes were plated onto micropatterned and planar PS surfaces coated with or without laminin at initial densities ranging from 15,000 to 40,000 cells per mL and maintained for 1, 3, and 7 days in MMEM. Applying laminin resulted in significant improvement in astrocyte adhesion and spreading of F-actin microfilaments and glial intermediate filaments on the polystyrene substrates in the direction of the grooves.[1] After 1 and 3 days, approximately three times as many cells adhered to laminin-coated substrates compared to those without laminin. Furthermore, the presence of laminin had a significant effect on the alignment of the astrocytes on the 10/20/3 μm patterned PS substrates. At initial plating densities ranging from 15,000 to 40,000 cells per mL, more than 85% alignment (≤ 20° with the direction of the grooves) was achieved on the laminin-coated micropatterned PS substrates.

Micropatterned PS substrates were then used to explore the effects of substrate topography and the adsorption of PLL and laminin on the proliferation and differentiation of the AHPCs. The AHPCs were originally isolated from the brains of adult Fischer 344 rats and maintained in growth medium supplemented with basic fibroblast growth factor (human recombinant bFGF; 20 ng/mL).[3,4] For *in vitro* analysis on PS substrates, the AHPCs were dissociated and plated on the micropatterned PS substrates coated with PLL/laminin at initial densities from approximately 20,000 to 40,000 cells per mL. These cells were maintained in the culture medium stated above without bFGF and having 1% v/v FBS for 3, 7, and 14 days. Under these conditions, the AHPCs selectively adhered and extended processes axially along the grooves of the patterned substrates owing to the effects of physical and chemical guidance mechanisms. Differentiation of the AHPCs on the micropatterned PS substrates was assessed morphologically and immunocytochemically up to 14 days *in vitro* using cell type–specific antibodies. After 7 days *in vitro*, approximately 50% of the

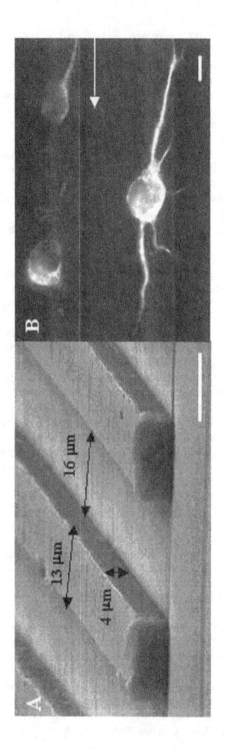

FIGURE 1. (**A**) Scanning electron micrograph image of a PS substrate having groove dimensions 16/13/4 μm created by solvent casting onto a silicon microdie created using microfabrication techniques. (**B**) AHPCs are aligned in the groove direction and are expressing the neuronal marker MAP-2ab (microtubule-associated protein 2ab) after 3 DIV on 16/13/4 μm micropatterned PS substrates in the presence of PLL and laminin. *Arrow* indicates groove direction. Scale bars = 10 μm.

AHPCs expressed a neuronal marker, MAP-2ab. Furthermore, nearly 75% of the GFP-expressing cells aligned (≤ 20° with the direction of the grooves) on the micro-patterned PS substrates with PLL/laminin-coated substrates. The AHPCs appeared highly elongated and extended their processes in the direction of the grooves and along groove boundaries. After 7 and 14 days, approximately 70% of the MAP-2ab IR cells were aligned in the direction of the grooves (FIG. 1B.). It appears that a combination of chemical (PLL/laminin) and physical (the micropatterned substrate) cues influence the directional guidance of astrocytes as well as neural progenitor cells.

To further explore the outgrowth and differentiation of the AHPCs, we have integrated the biological influence of astrocytes with physical and chemical guidance cues. In order to investigate the relationship between the degree of astrocyte alignment and AHPC outgrowth, AHPCs were cultured on top of near-confluent astrocyte monolayers formed on laminin-coated (0.01 mg/mL in EBSS) micropatterned PS substrates. The co-cultures were maintained for 3 days in a mixed medium that consisted of a 1:10 dilution of MMEM (1% v/v FBS) in a 1:1 mixture with AHPC differentiation media, resulting in a medium consisting of a final FBS concentration of 0.5%. The growth of AHPCs was enhanced by the presence of aligned astrocytes on the micropatterned substrate. On the patterned substrates, AHPC outgrowth was generally in the same direction of the aligned astrocytes. As the majority of the AHPCs were not in direct contact with the PS substrate, it appears that directed outgrowth may be stimulated by properties of the astrocyte surface. It was further observed that contact-mediated or soluble guidance cues provided by the astrocytes may have influenced AHPC differentiation.

Our results show that a synergistic combination of guidance cues enabled oriented growth of astrocytes and AHPCs and may influence terminal differentiation of the progenitor cells. The integration of the influence of the aligned astrocytes and the physical and chemical cues presented can potentially generate a permissive environment for specific cellular behavior such as the orientation and selective differentiation of neural stem/progenitor cells as well as directed neurite outgrowth during development and regeneration. The information gained from this integration of guidance cues has important applications in guided nerve regeneration *in vivo* within the central nervous system.

REFERENCES

1. RECKNOR, J.B. *et al.* 2004. Oriented astroglial cell growth on micropatterned polystyrene substrates. Biomaterials **25:** 2753–2767.
2. MILLER, C. *et al.* 2001. Oriented Schwann cell growth on micropatterned biodegradable polymer substrates. Biomaterials **22:** 1263–1269.
3. TAKAHASHI, J., T.D. PALMER & F.H. GAGE. 1999. Retinoic acid and neurotrophins collaborate to regulate neurogenesis in adult-derived neural stem cells cultures. J. Neurobiol. **38:** 65–81.
4. PALMER, T.D., J. TAKAHASHI & F.H. GAGE. 1997. The adult rat hippocampus contains primordial neural stem cells. Mol. Cell. Neurosci. **8:** 389–404.

Cellular and Molecular Regulation of Hematopoietic and Intestinal Stem Cell Behavior

XI C. HE,[a] JIWANG ZHANG,[a] AND LINHENG LI[a,b]

[a]*Stowers Institute for Medical Research, Kansas City, Missouri 64110, USA*

[b]*Department of Pathology, Kansas University Medical Center, Kansas City, Kansas 66160, USA*

ABSTRACT: Two fundamental questions in stem cell research are what controls stem cell number *in vivo* and which signal pathways regulate self-renewal. Here we summarize our recent studies regarding the role of BMP signaling in regulation of stem cell behavior in both the hematopoietic and intestinal systems. These studies provide evidence to show that BMP signaling plays an important role in controlling stem cell number, at least in these two stem cell compartments. However, the BMP signal utilizes different mechanisms to fulfill this purpose: in the hematopoietic stem cell compartment it controls stem cell number through regulation of the niche size; in the intestinal stem cell compartment it directly controls self-renewal of stem cells through restriction of Wnt/β-catenin activity. The *Bmpr1a* mutant mouse provided an elegant model which allowed us to identify the HSC niche, an enigma for more than 25 years. Our work provided more evidence to demonstrate the essential function of the niche in maintenance of stem cells and showed that multiple signals are required to maintain a balanced control of stem cell self-renewal.

KEYWORDS: hematopoiesis; intestines; polyposis; self-renewal; stem cells; niche; BMP; PTEN; Wnt; Akt

INTRODUCTION

Stem cells have the ability to self-renew and the potential to produce differentiated cells. Therefore, they play important roles in cellular specialization, pattern formation and organogenesis during embryogenesis, and in tissue regeneration in adults.[1] According to the stages of embryonic development, stem cells can be divided into four different types[2]: (1) the *totipotent stem cell* is the fertilized zygote and its immediate progeny; (2) the *pluripotent embryonic stem cell* (ESC), derived from an inner cell mass (ICM) of blastocysts. ESCs have the ability to give rise to all three embryonic germ layers—ectoderm, endoderm, and mesoderm—but not to trophoblast cells. ESCs then produce (3) *germline stem cells* for reproduction and (4)

Address for correspondence: Linheng Li, Stowers Institute for Medical Research, 1000 E. 50th Street, Kansas City, MO 64110. Voice: 816-926-4081; fax: 816-926-2023.
lil@stowers-institute.org

Ann. N.Y. Acad. Sci. 1049: 28–38 (2005). © 2005 New York Academy of Sciences.
doi: 10.1196/annals.1334.005

somatic stem cells, which form different types of adult stem cells. Unlike ESCs, adult stem cells have a more restricted developmental potency and therefore are multipotent. Embryonic stem cells and adult stem cells share the feature of self-renewal and therefore have been proposed to have "stemness."[3,4] Yet to be addressed is whether molecularly they use the same network or signal pathway to control self-renewal. ESCs have the advantage of being maintained and expanded *in vitro* under proper culture conditions, but it is still a huge task to induce ESCs into different types of tissue cells robustly. In contrast, tissue-specific adult stem cells naturally give rise to the corresponding tissues. However, the inability to maintain and expand most of adult stem cells *in vitro* significantly limits the mechanistic study of their regulation and hinders clinical applications. In order to be able to maintain and expand adult stem cells *in vitro*, understanding where and how adult stem cells are maintained *in vivo* will provide important insight into the mechanisms regarding regulation of stem cell self-renewal and uncover the signaling pathways involved in this regulation.

THE FUNCTION OF STEM CELL NICHE IN CONTROL OF STEM CELL PROPERTY

Bone marrow serves as the pioneer system for studying stem cells, and hematopoietic reconstitution of transplanted bone marrow in recipient mice provided the first evidence of stem cell existence.[5] Indeed, hematopoietic stem cells (HSCs) isolated from bone marrow are responsible for producing blood cells and supporting the immune system throughout life.[6] Only a few of the primitive self-renewing HSCs are able to generate donor-derived blood cells in the recipient animal and continue this process throughout its remaining lifespan.[7,8] Therefore, the donor-derived HSCs must progress through an expansion *in vivo* after homing in the host bone marrow. However, after transplantation into recipient mice, HSCs undergo limited expansion, resulting in a homeostatic number of stem cells. This suggests that there is limited physical space for HSCs *in vivo*. In 1978, Ray Schofield proposed a hypothesis that stem cells reside in a physically limited microenvironment, named niche, *in vivo*.[9] Marrow stromal cells and bone marrow–derived adherent cell lines can mimic the microenvironment *in vitro* to support prolonged proliferation and differentiation of stem cells and precursor cells, supporting this microenvironment model.[10–14] For the past four decades, identification of HSCs and studies regarding how HSCs give rise to progenitors, and further give rise to different types of myeloid and lymphoid lineage cells, have been well described. Ironically, the exact location, or niche, of HSCs in the bone marrow has been elusive until recently. This was mainly due to (1) lack of a unique marker to specifically recognize HSCs, (2) the complicated anatomic structure of bone marrow and morphologically indistinguishable hematopoietic lineage cells, and (3) technical difficulties with histologic sectioning of intact bone and bone marrow.

In principle, the stem cell niche can be identified by its unique anatomic position adjacent to the stem cell if the location of the stem cell is known. This has been exemplified by the identification of "cap" and "hub" cells as the niche for germline stem cells in *Drosophila* ovary and testis, respectively.[15–17] The physical location of germ stem cells (GSCs) in *Drosophila* ovary and testis is known owing to their well-defined anatomic structure, in which cells at different developmental stages are sequentially distributed. The lessons from studying GSCs and their microenvironment in flies led to the identification of the following features and functions regarding the stem cell niche:[18,19]

(1) The stem cell niche is composed of a special group of cells that provide a microenvironment for the maintenance of stem cells.

(2) The niche functions as a harbor where stem cells physically anchor and adherens molecules mediate this function.

(3) The niche generates key factors that regulate the balance between self-renewal and differentiation, thereby playing an essential role in controlling stem cell fate and maintaining the stem cell number, and ensuring a tight control of the stem cells so that tumorigenesis does not overpower ongoing differentiation.

In mammals, we propose that the niche also requires having additional functions:

(4) In most cases, adult stem cells attached to the niche are quiescent until they receive signals from the microenvironment that stimulate stem cell division. Upon division, one daughter cell is maintained in the niche as a stem cell (self-renewal); the other daughter cell leaves the niche, progresses through proliferation and differentiation, and eventually become functionally matured cells.

In our search for the HSC niche we started by locating the position of HSCs using HSC markers, including c-Kit and Sca-1, which have been successfully used for isolation of HSCs using flow cytometry.[20] However, this effort failed since neither c-Kit nor Sca-1 is specific for HSCs (Niu and Li, unpublished result). The reward of our endeavor in searching for the HSC niche came from studying a mutant mouse model in which we conditionally inactivated the bone morphogenetic protein receptor type 1a (Bmpr1a). The rationale behind our decision to genetically target the BMP signal pathway was due to the fact that the components of the BMP pathway are transcriptionally active in HSCs, as revealed by gene expression profiling analysis,[21,22] suggesting that BMP signaling may play a role in HSC regulation. The BMP (or Dpp) signal is essential in maintenance of GSCs in fly ovary.[23] Furthermore, BMPs have been shown to play important roles in induction of hematopoietic tissue during early embryonic development.[24,25] Still, the *in vivo* function of BMP signaling in the regulation of hematopoietic stem cells was unknown. In this *Bmpr1a* conditional mouse model, we found that the pool size of HSCs correlated with the volume of the trabecular bone.[26] Additionally, a correlation between HSC number and the volume of trabecular bone was observed in a simultaneous study using a transgenic mouse model in which parathyroid hormone receptor (PPR) was expressed.[27] Further analyses demonstrated that an increase in osteoblastic cells was responsible for the increase in HSC number in these two different mouse models.[26,27] Physically, we also found that HSCs attach to a subset of osteoblastic cells expressing high levels of N-cadherin, and importantly, N-cadherin and β-catenin form an adherens complex at the interface between HSCs and the osteoblastic cells.[26] Furthermore, the Jagged-Notch interaction was shown to be involved in osteoblast–HSC interaction[27] and has been demonstrated to be important for the maintenance of stem cells.[28–30] Taken together, these two studies (1) identified a specialized population of cells to which HSCs physically attached, (2) identified an N-cadherin and β-catenin adherens complex between HSCs and the osteoblastic cells, (3) showed that Jagged1 generated from osteoblast signaling through Notch influences HSCs, and(4) demonstrated that both the volume of trabecular bone and the number of N-cadherin$^+$ osteoblastic cells control the number of HSCs. All these observations fulfill the criteria for stem cell niche as discussed above. Thus, we concluded that the N-cadherin$^+$CD45$^-$ osteoblastic cells are a key component of niche supporting HSCs.[26,27] This conclusion is consistent with previous findings by several groups that *in vitro* co-culture of HSCs with osteoblasts can expand the HSC population.[31] Osteoblasts promote engraftment of allogeneic hematopoietic stem cell transplantation.[32] Recently, two reports provided additional evidence to support our conclusion that depletion of osteoblasts leads to loss of hematopoietic tissue and angiopoitin (Ang-1) and Tie-2 regulate HSC quiescence partially through N-cadherin.[33,34]

THE NICHE PROVIDES MULTIPLE SIGNALS TO IMPOSE A BALANCED CONTROL OF STEM CELL SELF-RENEWAL

In the previous section we discussed the role of BMP signaling in control of stem cell number through restriction of the niche size. However, this is an indirect mechanism and we questioned whether BMP signaling could also directly influence adult stem cell behavior. In the *Bmpr1a* mutant mice we targeted the HSC compartment as well as the compartments of intestinal and hair follicle stem cells. These three compartments are the most active stem cell compartments in the body and they are sensitive to the interferon signal. Since we utilized an interferon-inducible Cre-expression mouse line, we were thus able to look into all three stem cell compartments. We will now focus on the intestinal stem cell system.

The intestinal architecture is composed of the villus region with differentiating epithelial cells and the crypt region, where stem cells and proliferation cells are mainly located. Postnatal regeneration begins with intestinal stem cells (ISCs), which are located at the crypt bottom.[35] ISCs have the ability to regenerate themselves through self-renewal and are multipotent, giving rise to four different types of epithelial lineages: *columnar enterocytes* (the most abundant type) with apical microvilli lining the surface that function to absorb nutrition; *mucin-producing goblet cells* with plentiful mucus granules; *Paneth cells,* which secrete anti-microbial peptides and therefore contain large secreting granules; and *enteroendocrine cells,* which secret neuropeptides.[36] While Paneth cells are located at the crypt base, all other cell lineages are highly organized along the villus region; they differentiate during a rapid upward migration from crypt to the villus tip, and are finally shed into the lumen at the tip of villus. Cells shed into the lumen must be replaced by a steady reserve of cells generated in the crypts. During this continuous upward migration and maturation, the location of a cell within the migratory stream reveals its stage of development. Therefore, intestinal stem cells (ISCs), transient amplifying (TA) progenitors, functionally mature cells, and apoptotic cells are each confined to identifiable regions within each villus–crypt unit. This offers an attractive system for studying stem cell regulation and intestinal regeneration.[37]

Previously there was no unequivocal biological marker to recognize ISCs. The location of ISCs was identified at the 4th or 5th position of the crypt from the basement membrane by means of assay with BrdU- or ^{3}H-label.[39] This method has been used successfully to identify the stem cell location in hair follicles.[38] The label-retaining assay using BrdU or ^{3}H takes advantage of the fact that adult stem cells are, in most cases, in a quiescent state and therefore can retain the labeled BrdU or ^{3}H for a much longer time compared to cycling cells. In our study, we identified the ISC position using the BrdU-retaining (BrdU-R) method and further use the BrdU-R as the reference to examine signaling molecules in ISC and its niche cells.[40] To further define ISCs, we also showed that the BrdU-R cells located at this position are neither differentiated Paneth cells, which were recognized by presence of lysozyme,[40] nor highly proliferating cells, which can be identified with Ki67 (not shown). Since there are mainly three types of epithelial cells in the crypt—stem cells, proliferating progenitor cells, and differentiated Paneth cells—the BrdU-R cells located at this position are the best candidate for stem cells. Indeed, in regeneration experiments, it has been shown that a single surviving label-retaining cell in the crypt can regenerate the entire crypt after cytotoxic damage.[39] The clonal formation of the crypt–villus structure in the transgenic mouse model further supports the existence of functional stem cells in the crypt.[41,42] Taken together, this evidence supports the argument that the BrdU-R cells at the 4th or 5th position of the crypt from the basement membrane are ISCs.

Mutations in *BMPR1A* and *SMAD4* genes have been reported to cause juvenile polyposis syndrome (JPS) and juvenile intestinal polyposis (JIP), respectively, in humans,[43–45] indicating that the BMP pathway plays an essential role in regulation of intestinal development. Indeed, we found that the components of the BMP pathway are expressed in either ISC or its niche cells. BMP4 is expressed in the mesenchymal cells adjacent to ISCs, and therefore represents a niche signal. Bmpr1a is expressed on the surface of ISCs and is absent in proliferating cells. These expression patterns suggest that BMP4 generated from the niche can regulate stem cell property through its receptor, Bmpr1a. The active BMP signal is indeed present in ISCs, as confirmed by the presence of the phosphorylated (P) form of Smad1, 5, or 8. Interestingly, Noggin as a BMP4 antagonist, is also expressed in ISCs, but its expression is not constant, suggesting that transient expression of Noggin is able to regulate ISC behavior through inhibition of BMP signaling.[40]

Like mutations in *BMPR1a* and *SMAD4*, mutations in the *PTEN* (phosphatase homologue of tensin) gene can lead to Cowden disease [46] and Bannayan-Zonana syndrome (BRR),[47] both of which feature polyp development. This raises the possibility that the BMP and PTEN pathways interact genetically, and that PTEN is also involved in the regulation of intestinal homeostasis. PTEN (or TGFβ-regulated and epithelial cell–enriched phosphatase) is a tumor suppressor. Mutations in the PTEN gene have been found in a broad range of cancer, including breast, glioblastoma, endometrial, ovary, prostate, melanoma, skin, and gastrointestinal cancer.[48] The functions of PTEN include protein tyrosine phosphatase and lipid phosphatase activities.[49] One main function of PTEN is to antagonize PI3K activity, thus leading to inhibition of Akt (a Ser/Thr kinase), which is the main mediator of PI3K activity. Akt becomes activated following its phosphorylation in response to PI3K signaling. Thus, PTEN acts to inhibit PI3K/Akt activity, while the phosphorylated form of PTEN (P-PTEN) represents an inactivated PTEN.

Wnt signaling has been well characterized in the intestinal system and is known to promote progenitor cell proliferation, therefore favoring crypt cell fate.[50] It is also required for maintenance of ISCs, as inactivation of a Wnt-downstream molecule,

Tcf4, results in loss of crypt formation.[51] Wnt signals through the Frizzled receptor and is also controlled by a co-receptor LRP6 and a negative complex of APC/Axin/GSK3β. When Wnt is absent, APC controls the phosphorylation and the subsequent degradation of β-catenin through GSK3β. Active Wnt signaling leads to phosphorylation of LRP6, which subsequently results in inactivation of the APC–axin complex, thereby leading to translocation of β-catenin into the nucleus, where it binds its partner, TCF/LEF, to act on target genes.[52–55] Inappropriate activation of the Wnt signaling pathway causes stem/progenitor cells to overproliferate and finally results in adenomatous polyposis coli. [56–59]

Intriguingly, disruption in the BMP and PTEN pathways or abnormal activation of the Wnt pathway leads to diseases sharing a common feature of polyp development, albeit resulting in different types. This observation suggests that these signal pathways cross-talk as a regulatory network to regulate intestinal development and regeneration. Indeed, inactivation of *Bmpr1a* in mice results in polyp development throughout the intestinal tract in the mutant mice, a condition resembling juvenile polyposis syndrome (JPS) in humans.[40] A similar finding was also reported when BMP signaling was inhibited by overexpression of Noggin.[60] In both of these JPS models there was a significant expansion in the proliferation cell population and this was primarily due to an increase in the crypt number.[40,60] Since Wnt signaling is known to favor crypt cell fate and promote cell proliferation,[61] an increase in the crypt number implies that Wnt signaling is enhanced when BMP signaling is blocked. Indeed, this hypothesis was confirmed as measured by the activated (phosphorylated) form of LRP-6 or an increased nuclear localization of β-catenin.[40,60]

The question regarding how the BMP and Wnt signaling pathways interact still remained. To address this question, we decided to determine the Wnt-dependent transcriptional activity of β-catenin. While the active Wnt signal was found to be widely spread in all proliferating cells, including stem cells, as measured by P-LRP6, the transcriptional activity of β-catenin was primarily detected in a subset of cells including ISCs.[40] This observation led us to the hypothesis that Wnt signaling is required, but may not be sufficient to fully activate β-catenin in ISCs, and therefore a secondary signal was needed to cooperate with the Wnt signal in this regard. In searching for this secondary signal we found that (1) the inactivated (phosphorylated) form of PTEN was primarily present in ISCs; (2) Akt, as the downstream of PI3K which is subject to PTEN negative regulation, was also primarily present in ISCs as an active form; and (3) both the transcriptional activity and nuclear localization of β-catenin in ISCs are correlated with the presence of P-PTEN and P-Akt.[40] The transcriptional activity of β-catenin was determined by *in vitro* transfection assay, *ex vivo* electroporation assay, and *in vivo* Top-GAL activity measurement.[40] In these experiments we demonstrated that Akt is involved in the regulation of β-catenin activity, and this function may be required to coordinate with the Wnt signal to fully activate β-catenin in ISCs. However, the role of Akt in regulating β-catenin activity through GSK3β is debatable.[62,63] These controversial observations most possibly were due to differences in the assay systems. Indeed, Tian and colleagues recently reported that Akt is able to directly phosphorylate β-catenin and this phosphorylation primes 14-3-3ζ binding, thereby facilitating β-catenin stabilization, and this interaction seems to be limited to a small subset of crypt cells including ISCs.[64] Thus PTEN, as a negative regulator of PI3K/Akt signaling, is able to inhibit the transcriptional activity of β-catenin through suppression of Akt activity. This argument is supported by a previous report that PTEN, through inhibition of Akt, restricts nuclear accumulation of β-catenin.[65]

Now the question became: What signal regulates PTEN activity? BMP was reported to be able to enhance PTEN activity,[66] supporting the idea that the BMP and PTEN pathways may interact genetically.[67] To address this issue we carried out both *in vitro* and *in vivo* experiments and found that BMP positively regulates PTEN activity, but Noggin negatively regulates PTEN activity. Although the exact biochemical mechanism needs further investigation, we found that BMP signaling blocks the conversion of PTEN to P-PTEN, while Noggin favors P-PTEN formation.[40] We further showed that BMP signaling is able to inhibit β-catenin activity through regulation of PTEN activity; Noggin, on the other hand, is able to enhance β-catenin activity through suppression of BMP signaling.[40] This observation is consistent with other reports that overexpression of Noggin in Noggin-transgenic mice leads to enhanced Wnt signaling[60] and that Noggin–Wnt coordination is required for initiation of the hair growth cycle.[68]

The Wnt signal favors cell proliferation and promotes cell growth and therefore has to be restricted in adult stem cells to prevent tumorigenesis. Otherwise, abnormal activation of β-catenin leads to overproliferation of stem cells and results in tumors in both intestines and hair follicles.[50,69] The BMP signal generated from the ISC niche may serve as a "gatekeeper" to prevent abnormal activation of stem cells through suppression of Wnt/β-catenin activity. While we have shown that the PTEN-Akt pathway mediates the convergence of BMP and Wnt pathways to control β-catenin activity, we tend to favor the model that the BMP-Smad–mediated transcription regulation also contributes to the control of Wnt signaling as mutation in *Smad4* also causes juvenile polyposis.[45] This merits further investigation. Nevertheless, the ability of BMP signaling to inhibit Wnt/β-catenin activity in ISCs led us to the hypothesis that the BMP signal plays a role in inhibition of ISC self-renewal, thus balancing the positive control of self-renewal by Wnt/β-catenin signaling.[70] Indeed, a significant increase in the ISC number in the *Bmpr1a* mutant mice strongly supports this hypothesis. Finally, stem cells have to be activated to support ongoing intestinal regeneration. We propose that transient expression of Noggin may function as a "molecular switch" by temporally overriding the BMP restriction signal and coordinating with the Wnt signal to fully activate β-catenin, thus leading stem cells to activation for both self-renewal and tissue regeneration.

CONCLUSION

In this review, we have summarized our recent studies regarding the role of BMP signaling in regulation of stem cell behavior in both the hematopoietic and the intestinal systems. A common feature is that BMP signaling appears to play an important role in control of the stem cell number, at least in these two stem cell compartments. In the hematopoietic stem cell compartment the BMP signal controls the stem cell number indirectly through regulation of the niche size, and in the intestinal stem cell compartment it directly controls the self-renewal of stem cells through suppression of Wnt/β-catenin activity.

The *Bmpr1a* conditional KO line provided us with a great model to gain an insight into the dependence of stem cells on the microenvironment, and therefore

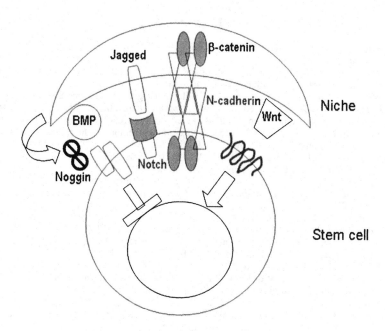

FIGURE 1. Basic features of the stem cell niche.

helped us to identify the HSC niche. Studies from our own work and from other groups in both flies and mammals have demonstrated the important function of the niche in supporting stem cells and further defined basic features of the stem cell niche as shown in FIGURE 1.

In summary, these are: (1) a specialized group of cells function as a key component of the stem cell niche; (2) N-cadherin and β-catenin form an adherens complex between stem cells and the niche, facilitating anchoring of stem cells to the niche; (3) the Jagged-Notch interaction between the niche-HSCs is important for the maintenance of stem cells as undifferentiated; (4) the niche-derived BMP signaling restricts stem cell activation and subsequent self-renewal through suppression of Wnt/β-catenin signaling, thus keeping a balance between self-renewal and tissue regeneration; and finally (5) Noggin, as an BMP antagonist, may play a role as a "molecular switch" to initiate activation of stem cells by temporally overriding the BMP restriction signal.

ACKNOWLEDGMENTS

We appreciate the proofreading and editing done by D. di Natale. We apologize to those whose papers are not cited here due to limited space; it does not mean their work is not significant. Our work is supported by the Stowers Institute for Medical Research.

REFERENCES

1. ROSSANT, J. 2004. *In* Handbook of Stem Cells. L. Lanza, *et al.*, Eds. Elsevier Academic Press. London.
2. WEISSMAN, I.L. 2000. Translating stem and progenitor cell biology to the clinic: barriers and opportunities. Science **287:** 1442–1446.
3. IVANOVA, N.B. *et al.* 2002. A stem cell molecular signature. Science **298:** 601–604.
4. RAMALHO-SANTOS, M., S. YOON, Y. MATSUZAKI, *et al.* 2002. "Stemness": transcriptional profiling of embryonic and adult stem cells. Science **298:** 597–600.
5. TILL, J.E. & E.A. MCCULLOCH. 1961. A direct measurement of the radiation sensitivity of normal mouse bone marrow cells. Radiat. Res. **14:** 213.
6. SPANGRUDE, G.J., S. HEIMFELD & I.L. WEISSMAN. 1988. Purification and characterization of mouse hematopoietic stem cells. Science **241:** 58–62.
7. DICK, J.E., M.C. MAGLI, D.A. HUSZAR, *et al.* 1985. Introduction of a selectable gene into primitive stem cells capable of long-term reconstitution of the hemopoietic system of W/Wv mice. Cell **42:** 71–79.
8. LEMISCHKA, I.R., D.H. RAULET & R.C. MULLIGAN. 1986. Developmental potential and dynamic behavior of hematopoietic stem cells. Cell **45:** 917–927.
9. SCHOFIELD, R. 1978. The relathionship between the spleen colony-forming cell and the hamatopopietic stem cell: a hypothesis. Blood Cells **4:** 7–25.
10. DEXTER, T.M., M.A. MOORE & A.P. SHERIDAN. 1977. Maintenance of hemopoietic stem cells and production of differentiated progeny in allogeneic and semiallogeneic bone marrow chimeras in vitro. J. Exp. Med. **145:** 1612–1616.
11. SITNICKA, E., F.W. RUSCETTI, G.V. PRIESTLEY, *et al.* 1996. Transforming growth factor beta 1 directly and reversibly inhibits the initial cell divisions of long-term repopulating hematopoietic stem cells. Blood **88:** 82–88.
12. RIOS, M. & D.A. WILLIAMS. 1990. Systematic analysis of the ability of stromal cell lines derived from different murine adult tissues to support maintenance of hematopoietic stem cells in vitro. J. Cell Physiol. **145:** 434–443.
13. MOORE, K.A., H. EMA & I.R. LEMISCHKA. 1997. In vitro maintenance of highly purified, transplantable hematopoietic stem cells. Blood **89:** 4337–4347.
14. ROECKLEIN, B.A. & B. TOROK-STORB. 1995. Functionally distinct human marrow stromal cell lines immortalized by transduction with the human papiloma virus E6/E7 genes. Blood **85:** 997–1005.
15. XIE, T. & A.C. SPRADLING. 2000. A niche maintaining germ line stem cells in the *Drosophila* ovary. Science **290:** 328–330.
16. KIGER, A.A., D.L. JONES, C. SCHULZ, *et al.* 2001. Stem cell self-renewal specified by JAK-STAT activation in response to a support cell cue. Science **294:** 2542–2545.
17. TULINA, N. & E. MATUNIS. 2001. Control of stem cell self-renewal in *Drosophila* spermatogenesis by JAK-STAT signaling. Science **294:** 2546–2549.
18. LIN, H.T. 2002. The stem-cell niche theory: lessons from flies. Nat. Rev. Genet. **3:** 931–940.
19. SPRADLING, A., D. DRUMMOND-BARBOSA & T. KAI. 2001. Stem cells find their niche. Nature **414:** 98–104.
20. WEISSMAN, I.L., D.J. ANDERSON & F. GAGE. 2001. Stem and progenitor cells: origins, phenotypes, lineage commitments, and transdifferentiations. Annu. Rev. Cell Dev. Biol. **17:** 387–403.
21. AKASHI, K. *et al.* 2003. Transcriptional accessibility for genes of multiple tissues and hematopoietic lineages is hierarchically controlled during early hematopoiesis. Blood **101:** 383–389.
22. PARK, I. *et al.*. 2002. Differential gene expression profiling of adult murine hematopoietic stem cells. Blood **99:**, 488–498.
23. XIE, T. & A.C. SPRADLING. 1998. Decapentaplegic is essential for the maintenance and division of germline stem cells in the *Drosophila* ovary. Cell **94:** 251–260.
24. DAVIDSON, A.J. & L.I. ZON. 2000. Turning mesoderm into blood: the formation of hematopoietic stem cells during embryogenesis. Curr. Top. Dev. Biol. **50:** 45–60.
25. BHATIA, M. *et al.* 1999. Bone morphogenetic proteins regulate the developmental program of human hematopoietic stem cells. J. Exp. Med. **189:** 1139–1148.

26. ZHANG, J. *et al.* 2003. Identification of the haematopoietic stem cell niche and control of the niche size. Nature **425:** 836–841.
27. CALVI, L. M. *et al.* 2003. Osteoblastic cells regulate the haematopoietic stem cell niche. Nature **425:** 841–846.
28. CRITTENDEN, S.L., D.S. BERNSTEIN, J.L. BACHORIK, *et al.* 2002. A conserved RNA-binding protein controls germline stem cells in *Caenorhabditis elegans*. Nature **417:** 660–663.
29. VARNUM-FINNEY, B. *et al.* 1998. The Notch ligand, Jagged-1, influences the development of primitive hematopoietic precursor cells. Blood **91:** 4084–4091.
30. VARNUM-FINNEY, B. *et al.* 2000. Pluripotent, cytokine-dependent, hematopoietic stem cells are immortalized by constitutive Notch1 signaling. Nat. Med. **6:** 1278–1281 <taf/DynaPagetaf?file=/nm/journal/v6/n11/full/nm1100_1278html taf/DynaPagetaf?file=/nm/ournal/v6/n11/abs/nm1100_1278html>.
31. TAICHMAN, R.S. & S.G. EMERSON. 1998. The role of osteoblasts in the hematopoietic microenvironment. Stem Cells **16:** 7–15.
32. EL-BADRI, N.S., B.Y. WANG, CHERRY & R.A. GOOD. 1998. Osteoblasts promote engraftment of allogeneic hematopoietic stem cells. Exp. Hematol. **26:** 110–116.
33. VISNJIC, D. *et al.* 2004. Hematopoiesis is severely altered in mice with an induced osteoblast deficiency. Blood **103:** 3258–3264.
34. ARAI, F. *et al.* 2004. Tie2/angiopoietin-1 signaling regulates hematopoietic stem cell quiescence in the bone marrow niche. Cell **118:** 149–161.
35. BOOTH, C. & C.S. POTTEN. 2000. Gut instincts: thoughts on intestinal epithelial stem cells. J. Clin. Invest. **105:** 1493–1499.
36. BJERKNES, M. & H. CHENG. 1999. Clonal analysis of mouse intestinal epithelial progenitors. Gastroenterology **116:** 7–14.
37. HERMISTON, M.L. &J.I. GORDON. 1995. Organization of the crypt-villus axis and evolution of its stem cell hierarchy during intestinal development. Am. J. Physiol. **268:** G813–822.
38. COTSARELIS, G., T.T. SUN & R.M. LAVKER. 1990. Label-retaining cells reside in the bulge area of pilosebaceous unit: implications for follicular stem cells, hair cycle, and skin carcinogenesis. Cell **61:** 1329–1337.
39. POTTEN, C.S. & J.H. HENDRY. 1995. Radiation and Gut. C.S. Potten & J. Henry, Eds. Elsevier. Amsterdam.
40. HE, X.C. *et al.* 2004. BMP signaling inhibits intestinal stem cell self-renewal through suppression of Wnt-beta-catenin signaling. Nat. Genet. **36:** 1117–1121.
41. ROTH, K.A., M.L. HERMISTON & J.I. GORDON. 1991. Use of transgenic mice to infer the biological properties of small intestinal stem cells and to examine the lineage relationships of their descendants. Proc. Natl. Acad. Sci. USA **88:** 9407–9411.
42. GORDON, J.I., G.H. SCHMIDT & K.A. ROTH. 1992. Studies of intestinal stem cells using normal, chimeric, and transgenic mice. FASEB J. **6:** 3039–3050.
43. HOWE, J.R. *et al.* 2001. Germline mutations of the gene encoding bone morphogenetic protein receptor 1A in juvenile polyposis. Nat. Genet. **28:** 184–187.
44. HOWE, J.R. *et al.* 1998. A gene for familial juvenile polyposis maps to chromosome 18q21.1. Am. J. Hum. Genet. **62:** 1129–1136.
45. HOWE, J.R. *et al.* Mutations in the SMAD4/DPC4 gene in juvenile polyposis. Science **280:** 1086–1088.
46. LIAW, D. *et al.* 1997. Germline mutations of the PTEN gene in Cowden disease, an inherited breast and thyroid cancer syndrome. Nat. Genet. **16:** 64–67.
47. MARSH, D. J. *et al.* 1997. Germline mutations in PTEN are present in Bannayan-Zonana syndrome. Nat. Genet. **16:** 333–334.
48. MUTTER, G.L. 2001. Pten, a protean tumor suppressor. Am. J. Pathol. **158,** 1895–1898.
49. WU, H., V. GOEL & F.G. HALUSKA. 2003. PTEN signaling pathways in melanoma. Oncogene **22:** 3113–3122.
50. SANCHO, E., E. BATLLE & H. CLEVER. 2004. Signaling pathways in intestinal development and cancer. Annu. Rev. Cell Dev. Biol. **20:** 695–723.
51. KORINEK, V. *et al.* 1998. Depletion of epithelial stem-cell compartments in the small intestine of mice lacking Tcf-4. Nat. Genet. **19:** 379–383.
52. SAHL, W.J., JR. & H. CLEVER. 1994. Cutaneous scars: Part I. Int. J. Dermatol. **33:** 681–691.

53. SAHL, W.J., Jr. & H. CLEVER. 1994. Cutaneous scars: Part II. Int. J. Dermatol. **33:** 763–769.
54. NUSSE, R. *et al.* 1997. Cell culture and whole animal approaches to understanding signaling by Wnt proteins in *Drosophila.* Cold Spring Harbor Symp. Quant. Biol. **62:** 185–190.
55. PEIFER, M. & P. POLAKIS. 2000. Wnt signaling in oncogenesis and embryogenesis: a look outside the nucleus. Science **287:** 1606–1609.
56. MIYOSHI, Y. *et al.* 1992. Germ-line mutations of the APC gene in 53 familial adenomatous polyposis patients. Proc. Natl. Acad. Sci. USA **89:** 4452–4456.
57. JOSLYN, G. *et al.* 1991. Identification of deletion mutations and three new genes at the familial polyposis locus. Cell **66:** 601–613.
58. GRODEN, J. *et al.* 1991. Identification and characterization of the familial adenomatous polyposis coli gene. Cell **66:** 589–600.
59. NAKAMURA, Y. *et al.* 1991. Mutations of the adenomatous polyposis coli gene in familial polyposis coli patients and sporadic colorectal tumors. Princess Takamatsu Symp. **22:** 285–922.
60. HARAMIS, A.P. *et al.* 2004. De novo crypt formation and juvenile polyposis on BMP inhibition in mouse intestine. Science **303:** 1684–1686.
61. VAN DE WETERING, M. *et al.* 2002. The beta-catenin/TCF-4 complex imposes a crypt progenitor phenotype on colorectal cancer cells. Cell **111:** 241–250.
62. FUKUMOTO, S. *et al.* 2001. Akt participation in the Wnt signaling pathway through Dishevelled. J. Biol. Chem. **276:** 17479–17483.
63. DING, V.W., R.H. CHEN & F. MCCORMICK. 2000. Differential regulation of glycogen synthase kinase 3beta by insulin and Wnt signaling. J. Biol. Chem. **275:** 32475–32481.
64. TIAN, Q. *et al.* 2004. Proteomic analysis identifies that 14-3-3{zeta} interacts with {beta}-catenin and facilitates its activation by Akt. Proc. Natl. Acad. Sci. USA **101:** 15370–15375.
65. PERSAD, S., A.A. TROUSSARD, T.R. MCPHEE, *et al.* 2001. Tumor suppressor PTEN inhibits nuclear accumulation of beta-catenin and T cell/lymphoid enhancer factor 1-mediated transcriptional activation. J. Cell Biol. **153:** 1161–1174.
66. WAITE, K.A. & C. ENG. 2003. BMP2 exposure results in decreased PTEN protein degradation and increased PTEN levels. Hum. Mol. Genet. **12:** 679–684.
67. WAITE, K.A. & C. ENG. 2003. From developmental disorder to heritable cancer: it's all in the BMP/TGF-beta family. Nat. Rev. Genet. **4:** 763–773.
68. JAMORA, C., R. DASGUPTA, P. KOCIENIEWSKI & E. FUCHS. 2003. Links between signal transduction, transcription and adhesion in epithelial bud development. Nature **422:** 317–322.
69. GAT, U., R. DASGUPTA, L. DEGENSTEIN & E. FUCHS. 1998. De novo hair follicle morphogenesis and hair tumors in mice expressing a truncated beta-catenin in skin. Cell **95:** 605–614.
70. REYA, T. *et al.* 2003. A role for Wnt signalling in self-renewal of haematopoietic stem cells. Nature **423:** 409–414.

The Function of the Neuronal Proteins Shc and Huntingtin in Stem Cells and Neurons

Pharmacologic Exploitation for Human Brain Diseases

CHIARA ZUCCATO,[a] LUCIANO CONTI,[a] ERIKA REITANO, MARZIA TARTARI, AND ELENA CATTANEO

Department of Pharmacological Sciences and Center of Excellence on Neurodegenerative Diseases, University of Milan, Milan, Italy

ABSTRACT: The identification of intracellular molecules and soluble factors that are important for neuronal differentiation and survival are of critical importance for development of therapeutic strategies for brain diseases. First, the activity of these factors/molecules may be enhanced *in vivo* in the attempt to induce proper neuronal differentiation and integration of the resident stem cells. Second, these factors may be applied *ex vivo* to increase the recovery of neurons from stem cells. Third, for those intracellular molecules that play crucial roles in neuronal survival, identification of their downstream targets may give us the chance to develop drug screening assays that use these targets for therapeutic purposes. In recent years, it has become evident that intracellular signaling processes are critical mediators of the responses of neural stem cells and neurons to growth factors. Analysis of the mechanisms of signal transduction has led to the striking finding that a handful of conserved signaling pathways appear to be used in different combinations to specify a wide variety of tissues or cells. This review will focus on the mechanisms by which specific molecules control the transition from proliferation to differentiation of neural progenitor cells and the subsequent survival of postmitotic neurons; it also discusses how this knowledge may be exploited to increase the potential efficacy of stem cell replacement in the damaged brain.

KEYWORDS: shc molecules; signal transduction; neural stem cells; mature neurons; neuronal survival; neurodegenerative diseases; huntingtin; Huntington's disease

NEURAL STEM CELLS AS A TOOL FOR BRAIN REPAIR: AN INTRODUCTION

Throughout the last decade there has been an increasing interest in the study of the biology of stem cells. These are widely considered as an invaluable potential tool

[a]These authors contributed equally to this manuscript and should be considered co–first authors.

Address for correspondence: Elena Cattaneo, Department of Pharmacological Sciences and Center of Excellence on Neurodegenerative Diseases, University of Milan, Milan, Italy. Voice: +39-02-5031 8333; fax:+39-02-5031 8284.

Elena.Cattaneo@unimi.it

Ann. N.Y. Acad. Sci. 1049: 39–50 (2005). © 2005 New York Academy of Sciences.
doi: 10.1196/annals.1334.006

for cell therapy approaches to a broad range of clinical conditions. The rationale of this approach relies on transplantation to the central nervous system of cells that are able to replace the lost elements and rewire the disrupted circuitries, thus restoring normal brain function.[1] Since mature neurons do not survive dissection and grafting procedures, immature or proliferating cells represent the best source of donor cells for transplantation to the central nervous system (CNS).

According to its definition, the main physiologic function of a stem cell is to generate all of the differentiated cell types of the tissue in which it resides. Indeed, a stem cell is generally defined operationally as a cell that is: (i) multipotent, (ii) capable of self-renewal, and (iii) capable of generating a progeny that can functionally integrate into and repair the tissue of origin. This implies that the progeny of neural stem cells (NCSs)—that is, those stem cells residing inside the nervous system—will include mature neurons, astrocytes, and oligodendrocytes. Stem cell technology is particularly important for the central nervous system (CNS) since cell transplantation might help to overcome the intrinsic poor capability of the nervous tissue to replace elements lost in the course of injury or disease. We believe, however, that this cannot preclude pharmacological treatments aimed at preserving the activity of the undamaged neurons and the survival of the donor cells. Proper control over the differentiation and integration pattern of brain stem cells may therefore eventually allow the treatment of a wide range of degenerative diseases characterized by neuronal or glial loss. At this stage it is therefore important to increase our knowledge of the molecular regulators and genetic cascades that control NSCs self-renewal and multipotency as well as proteins whose activity may be enhanced pharmacologically to increase neuroprotection in the brain.

NEURAL STEM CELLS: SOURCES, PROPERTIES, AND MISSING ELEMENTS

The demonstration of the existence of cells in the developing and adult nervous system that can be isolated and grown *in vitro* where they behave as *bona fide* neural stem cells [2] sustains the hopes of scientists and patients. However, toward such a goal, and in spite of the excitement, a great deal of basic research is still needed since the *in vitro* properties of these cells remain poorly understood and their *in vivo* engraftment capacity is of limited efficacy. During brain development, NSCs are localized in the epithelial layer of the germinal zone surrounding the ventricles.[3] As brain maturation continues, postmitotic neurons migrate away from the ventricular zone, mainly guided by radially oriented glial processes, and the ventricular zone diminishes in size.[4] In the adult brain, cells with similar stem-like properties also exist, mostly originating from two regions: the subventricular zone (SVZ) of the lateral ventricles and the hippocampus.[2] It is noteworthy that different studies indicate that NSCs from different fetal and adult brain areas are not identical, as demonstrated by different growth characteristics, trophic factor requirements, and specific patterns of differentiation.[3] *In vitro* studies indicate that two types of neural stem-like cells can be isolated in multiple brain regions and appear to coexist. One type shows epidermal growth factor (EGF) responsiveness and can be expanded as floating cell aggregates, called neurospheres,[5] becoming fibroblast growth factor (FGF)–responsive with the *in vitro* passages.[6,7] The second group has been shown to be FGF-dependent and can be propagated both as adherent cultures as well as neurospheres.[8] Nevertheless, a major lack in the field, is the absence of efficient and validated procedures for the prospective identification and isolation of the brain stem cells. Indeed, all the above-mentioned evidence indicating the existence of heterogeneous NSC populations is the consequence of the lack of a neural stem–restricted marker for NSC prospective isolation. This implies that it is currently very difficult to distinguish, but posteriorly and only following accurate clonal analysis, between a real NSC and a progenitor. Conceptually, these two populations differ in their differentiative capabilities. Indeed, while NSCs are multipotential, brain progenitors are considered to be more limited in their potential and able to produce only restricted phenotypes.[9] Up to date, there are only few markers of putative NSCs that may be used with some degree of specificity that are expressed by brain progenitors and not by other cell types.[1] In the absence of a clear *in vivo* assay to identify the NSCs, most authors use the ability to grow, *in vitro*, as neurospheres which contain cells capable of differentiating as glia and neurons, as an operative definition for "NSCs."[1] Neurospheres are, however, cell aggregates that grow in suspension and are composed by a heterogeneous population of cells at different maturation stages (from *bona fide* stem to progenitors, to mature and immature neurons and glia) and that are capable of differentiating *in vitro* as glia and neurons, even after several passages. Estimation of the number of *bona fide* stem cells contained in a preparation of cells dissociated from neurospheres varies as widely as for those contained in the fetal brain.[1,3,10] This reflects on the experimental program. Given the mixed nature of the cells in culture, their transplantation or gene expression profiling has produced contradictory results and is of limited impact.[1] New strategies to allow efficient prospective isolation of stem cells from the brain are highly necessary to guarantee uniformity of results and lead to trustworthy conclusions.

Outside the CNS, another potential source of NSCs is the blastocyst, from which rodent and human ES cells have been derived.[11] The power of ES cells lies in their more homogenous propagation and in the possibility of enriching the undifferentiated stem-cell-like population by means of stable genetic modifications. Genes that act as crucial markers of a stem cell state have been identified and the ability of these cells to generate a consistent number of different neuronal subtypes, even after extensive passages *in vitro* has been reported. Nevertheless, two critical issues remain unsolved with respect to the application of such cells in human brain transplantation: (i) whether the employment of the supernumerary frozen blastocysts derived from *in vitro* fertilization (IVF) procedures and that remain unused for several years, until they are physically discarded, should be allowed; and (ii) the likelihood that implanted ES cells could form tumors, although strategies to tackle this point are currently being devised.

Apart from the source of NSCs employed, another important requirement is for procedures that allow consistent and more uniform neuronal differentiation of the donor cells after intracerebral transplantation. The data available indeed demonstrate that cells derived from neurospheres and implanted into the adult or developing brain generate mostly glial cells with rare neuronal cells, when present (about 3–5%), suggesting that the poor control on the differentiative fate remains the main open question.[1] In this view, the development of strategies aimed at improving the rate of neuronal, over glial, differentiation and survival by genetically engineering NSCs represents one of the major challenge for the possible employment of these cells in cell replacement therapy. Such attempts and their success depend on a deeper under-

standing of the function of candidate genes/proteins or signaling pathways in order
to stimulate proper division and differentiation of NSCs.

SIGNALING MOLECULES INSTRUCTING NSC SELF-RENEWAL
AND NEURONAL DIFFERENTIATION

The role of several growth factors in the control of the proliferation, survival, and
differentiation of NSCs has been deeply investigated. Particularly intriguing are the
data showing that growth factors like EGF or bFGF can be mitogenic for the immature neural stem/progenitor cells, but act as survival and pro-differentiative agents
for postmitotic neurons. Among the possible explanations for these dual effects, it
has recently been proposed to be due to a developmental switch in the availability of
intracellular signaling proteins at the transition from stem/progenitor cell proliferation to differentiation.[12] Our studies have identified the regulated expression and activity of Shc(s) adapter molecules, which couple the signal from the activated

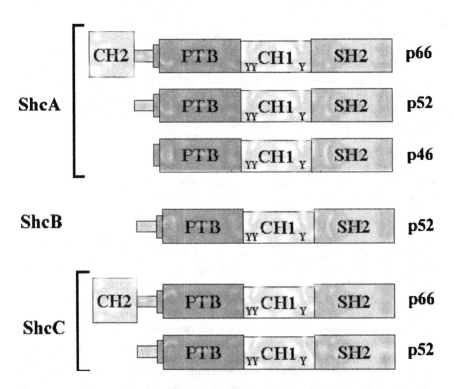

FIGURE 1. Schematic diagram of the structure of Shc molecules. Shc adaptors are
modular proteins characterized by an SH2 domain at the C-terminus, a central proline- and
glycine-rich region (CH1, collagen-homologous region 1) and a phosphotyrosine-binding
domain (PTB) N-terminus. The CH1 regions in ShcA, ShcB, and ShcC contain the three
tyrosine phosphorylation sites at Tyr239–240 and Tyr317.

receptor to the downstream effectors as a "simple" mechanism that the cell has developed to generate diversity in biological responses to growth factors (FIG. 1). Shc proteins, indeed appear to play a central role in the control of the proliferation and maturation of mitotically active neural stem/progenitor cells into postmitotic neurons.[12,13]

The Shc Family of Adaptor Molecules

Up to date, three Shc genes have been identified—ShcA, ShcB/Sli, and ShcC/ Rai/N-Shc—with a consistent homology.[12] These three Shc genes are characterized by the presence of phosphotyrosine regulatory residues and by the PTB (phosphotyrosine binding domain), CH1 (collagene homology 1, a proline-rich domain), and SH2 (Src-homology 2 domain) motifs in the presented order. Three isoforms are known for ShcA (of 66-, 52-, and 46-kDa), one isoform for ShcB (of 52-kDa), and two for ShcC (of 54- and 69-kDa).The p66^{ShcA} isoform displays a further N-terminal CH domain (CH2) that contains important regulatory serine residues. They share elevated homology in both the C terminus SH2 domain and the N terminus PTB domain, the most divergent sequence being in the proline- and glycine-rich CH1 (collagene homology 1) region. ShcA proteins have been extensively characterized and shown to be widely expressed outside the CNS. Their importance is indicated by (I) the early embryonic lethal phenotype of p52^{ShcA} null mutation;[14] (ii) the impairment in thymocyte development in conditional p52^{ShcA} knockout,[15] and (iii) by the increase in life span and resistance to stress stimuli in p66^{ShcA} knockout animals.[16]

Regulated Expression of ShcA and ShcC during Brain Maturation

Despite the apparently constitutive presence of ShcA in extraneural tissues, ShcA expression and activity within the brain appears to be tightly regulated during development and maximal at the early developmental stages (FIG. 2). Particularly, it has been previously demonstrated that ShcA proteins and mRNAs are sharply downregulated in coincidence with neurogenesis in the brain.[17] *In situ* hybridization analyses indicated that at day E10.5 the whole neural tube expresses high levels of the ShcA transcript. However, at later developmental stages (i.e., E12.5 embryos), the ShcA messenger RNA was concentrated in the germinal zone, but was found to be strongly reduced in the postmitotic areas, further reinforcing the idea of a link between division in the brain and expression of this adaptor molecule. These data were further confirmed by *in vivo* bromodeoxyuridine (BrdU) labeling of dividing stem/progenitor cells, which showed co-localization of the ShcA messenger RNA with the BrdU-immunostained cells. This co-localization was also confirmed for single isoform transcripts such as the less abundant, at this stage, p66^{ShcA} isoform. In the same way, the adult brain exhibited low ShcA expression, the main exception being the olfactory epithelium the subventricular zone.[18] These changes in the expression and activity of ShcA as a function of neuronal maturation were confirmed *in vitro* in differentiating neuronal cultures.[17] It was also found that, *in vivo*, immunoprecipitation of ShcA from the telencephalic vesicles of embryonic brains injected intraventricularly with mitogens (EGF) revealed a higher phosphorylation of the p52^{ShcA} isoform with respect to vehicle-injected control animals.[17] Particularly, in treated samples, Grb2 co-immunoprecipitation was observed, indicating that ShcA is not only present in the germinal epithelium, but it is also able to elicit a functional response to mitogens by recruitment of Grb2 and activation of the downstream Ras-MAPK pathway.

The demonstration that ShcA availability is tightly regulated and that, particularly, ShcA presence becomes limited during neural stem/progenitor cells maturation *in vivo* and *in vitro* leads to the proposition that these changes may influence as well the activity of the Ras-MAPK pathway during development and/or that other Shc-like proteins may substitute for ShcA function in mature neurons.[12] Given the existence of two new Shc members, ShcB and ShcC, the latter being selectively expressed in the brain, it has recently been suggested that one or both of them could replace ShcA in mature neurons.[19,20,21] Analyses of ShcC expression indeed showed that ShcA is replaced by ShcC, which is not expressed in neural progenitors, but it appears in early postmitotic neurons and reaches maximal levels in the adult brain, where it is found localized only in neurons. These data were further extended by

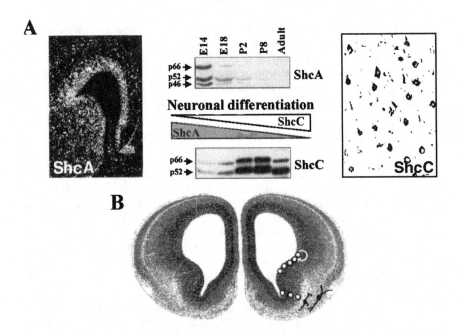

FIGURE 2. (**A**) Expression of ShcA/C during forebrain maturation. *Left panel:* Expression of ShcA mRNAs coincides with the proliferative epithelium. *In situ* analysis performed on embryonic day 14.5 brain section. (From Conti *et al.*[17] Reprinted by permission). *Middle panel:* Western blot analyses showing maximal expression of the three ShcA isoforms during early stages of neurogenesis, while ShcC expression increases at late maturation stages. (From Conti *et al.*[13] Reprinted by permission.) *Right panel:* ShcC is expressed in mature neurons in the adult rodent brain. (From Conti *et al.*[13] Reprinted by permission.) Immunohistochemistry performed on adult cortex section. (**B**) Schematic drawing: ShcA and ShcC in the embryonic brain. Schematic coronal section of the embryonic rodent brain recapitulating the expression of ShcA on proliferating neural stem cells lining the ventricles (*open circles* indicate stem cells that are ShcA-positive) and ShcC expression in migrating/postmitotic neurons (*black drawings* indicate postmitotic neurons that are ShcC-positive).

FACS analyses performed on a suspension of primary CNS cells isolated from embryonic day 14 (E14) embryos (when the stem/progenitor cells of the brain still actively divide) following a 30-minute *in vivo* exposure to BrDU. This analysis has confirmed that the BrdU-positive cells are ShcC-negative. Cells positive for ShcC are instead detectable by FACS analyses at later developmental stages and increase over time during embryonic and postnatal maturation of the brain. An immunohistochemical study at adult stages further revealed that ShcC is restricted to mature neurons. Various neuronal cell types in different brain regions were immunoreactive, indicating that ShcC is not restricted to particular subpopulations of CNS neurons. Importantly, no ShcC was detected in glial cells. These data have also been confirmed *in vitro* in cultures of freshly dissociated CNS primary cells induced to differentiate into neurons by exposure to chemically defined serum-free media. Similar changes in ShcA and ShcC levels during neuronal maturation have been observed in several mammalian species (rat, mouse and human).[13] The neuron-restricted expression of ShcC may be predictive of a key general role played in these cells. Particularly, given the above-described central roles of ShcA in signal transduction, ShcC appearance in differentiating NSCs has been hypothesized to serve different "connector functions" compared with ShcA.[13] In this regard, Pawson and colleagues[14] demonstrated the existence of a strict link between Shc levels and cell responsiveness. The authors showed that ShcA expression and activity are required in cells of the cardiovascular system to make them responsive to low concentrations of growth factors. Indeed, while a low concentration of growth factors is necessary to activate the MAPK pathway in mouse embryo fibroblasts (MEF), cells from ShcA knockout mice require a higher concentration of growth factors to activate the same signaling cascade. Transfection experiments in primary NSCs and in postmitotic neurons revealed that ShcC acts to promote neuronal differentiation and improve survival of these cells.[13]

Different Signals from Similar Molecules

Although ShcA and ShcC share a common architecture and a very close sequence similarity, they have been shown to differently modulate the signals from the activated tyrosine kinase receptors (FIG. 3). Biochemical studies indicate that ShcC elicits different effects from those of ShcA through a different kinetic of activation of downstream effector molecules. Indeed, ShcC was found to elicit neuronal differentiation via prolonged stimulation of the MAPK.[13,22] This behavior is reminiscent of that described in PC12 cells exposed to NGF, where persistent activation of MAPK is required for neuronal differentiation. On the contrary, ShcC-driven pro-survival effect occurs via recruitment of the PI3K-Akt pathway,[13,22] as demonstrated by the fact that its pharmacological or molecular inhibition markedly abolishes this effect. In this respect, ShcC-induced Akt activation was found to cause phosphorylation (with inhibition) of Bad, a proapoptotic member of the Bcl2 family.[13] Single and double ShcB/C null mice have been recently described.[23] ShcB-deficient mice exhibit a loss of peptidergic and nonpeptidergic nociceptive sensory neurons. ShcC null mice appear not to show gross anatomic abnormalities. It is noteworthy that mice lacking both ShcB and ShcC exhibit a significant additional loss of neurons within the superior cervical ganglia. This aspect may emphasize that the lack of phenotype in ShcC null mice could be due to a partial compensation by the other ShcB

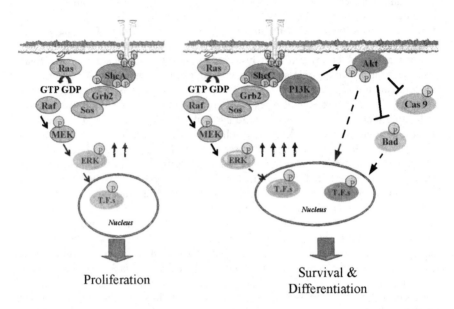

FIGURE 3. Different actions of ShcA and ShcC in the activation of Ras-MAPK and PI3k-Akt pathways.

or other Shc members during development, thus masking ShcC real function in neural tissues. Further analyses will be required to elucidate ShcC role *in vivo* in the mice.

Taken together, these results unveil a new scenario within which physiological changes in the availability of ShcA and ShcC adaptors during brain development act to modify neural stem/progenitor cell responsiveness as a function of the new and developing environment. Importantly, we believe that defining the specific requirements for ShcA or ShcC in the division and differentiation of neural stem cells and survival of neurons may provide a potential outcome in order to increase proliferation and expansion of neural precursors or their overall survival and neuronal differentiation. These proteins may also become interesting targets for drug screening approaches aimed at identifying drugs able to increase the survival and generation of neurons from stem cells.[24]

PHARMACOLOGIC EXPLOITATION OF NEURAL STEM CELLS IN NEURODEGENERATIVE DISEASES: THE IMPORTANCE OF HUNTINGTIN FUNCTION FOR THE TREATMENT OF HUNTINGTON'S DISEASE

Transplantation of stem cells and their neural derivatives within the adult brain has been proposed as future therapies for treating many diseases of the central nervous system. These include Parkinson's disease (PD), Huntington's disease (HD), amyotrophic lateral sclerosis(ALS), and Alzheimer's disease (AD). Many of these diseases are neurodegenerative disorders that are caused by specific neuronal populations either dying or beginning to malfunction. They are difficult to treat by means of stem cell transplantation because generally the function previously performed by neurons is permanently lost, and replacing cells lost through disease, considering the complexity of the human brain structure and function and the precise need for pattern repair and regulated activity of many neurotransmitters, is a very difficult goal to attain. Because of this limitation, diseases in which a specific area of the brain is damaged are the prime candidates for neural stem cell treatment. One example is given by Huntington's disease, an inherited neurological disorder caused by a defect in a single gene. The genetic cause of the disease is an excessive repeating of the CAG trinucleotide in exon 1 of the *huntingtin* gene. The result is the production of a protein bearing a polyglutamine expanded tract in its N-terminus which is toxic and leads to neurodegeneration of medium spiny projection neurons in striatum. As a consequence, patients with Huntington's disease exhibit motor dysfunction, cognitive decline, and psychiatric disturbance.[25] Discovery of the disease-linked gene in 1993 has made accurate diagnosis possible, but the disease still remains untreatable since its pathogenetic mechanism has been not completely elucidated and no drugs are available to cure or to slow the progression of the pathology. In this dramatic situation for patients, one important avenue that is already under testing for years is the use of cell therapy to replace dying striatal neurons. Transplantation of fetal progenitors has already been proposed for HD patients and has been shown to be of benefit at early stages, and in a small percentage of cases the graft was able to produce cognitive and motor improvements.[26] Other groups have indicated the need for larger clinical trials, which are presently under way in Europe.[27,28] Although small clinical trials have been carried out, transplanted fetal neurons survived without typical pathology: they developed striatal projection neu-

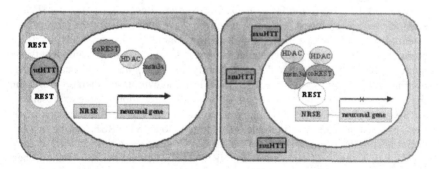

FIGURE 4. Huntingtin activity at the NRSE silencer involves the availability of the transcriptional repressor REST/NRSF (repressor element-1 transcription factor/neuron restrictive silencer factor). *Left:* Wild-type huntingtin binds REST in the cytoplasm. No corepressor complex is formed on the NRSE silencer, and neuronal gene transcription can proceed. *Right:* In HD, mutant huntingtin is less capable of sequestering REST in the cytoplasm. The NRSE silencer is therefore activated and transcription of NRSE-regulated genes (including the BDNF) is repressed.

rons and interneurons and received afferents from the patient's brain. However, analyses of the function of the huntingtin gene has shown that cell therapy might be not sufficient to cure HD and it should be combined with other strategies. Recent findings show that wild-type huntingtin gene product is essential for the production and delivery of brain-derived neurotrophic factor (BDNF) from the cortex to the target striatal neurons.[29] BDNF is a neurotrophin essential for striatal neuronal survival, and loss of BDNF cortical support to striatum contributes to striatal neurodegeneration, as recently demonstrated in mice in which cortical BDNF has been depleted via genetic manipulation.[30] Importantly, reduced cortical BDNF is found in HD, thus leading to a reduced supply of cortical BDNF to striatum. [29] In this case, cell replacement therapy may be less efficient given that the transplanted cells will be exposed to reduced trophic factor supply from the cortical afferents. Direct infusion of BDNF to the striatal target may be one strategy to increase stem cell efficacy in HD. An additional strategy may rely on the exploitation of the mechanisms through which loss of cortical BDNF production occurs in HD. The identification of such a mechanism might provide information for novel approaches to be combined to cell replacement which are based on the increase of cortical BDNF production. This means that in coincidence with the study of how to generate striatal projection neurons from neural stem cells and how to improve their transplantation integration and survival, basic research should also consider the study of disease-linked mechanisms. HD is an example of how this combined approach may become exploited. A recent study from our group shows that the different ability of wild-type and mutant huntingtin to modulate BDNF production mostly relies on a different capability of the two proteins to activate transcription from a region of the BDNF promoter that includes promoter-exon II.[29,31] The BDNF promoter is indeed composed of four 5 exons linked to separate promoters (I, II, III, IV), which are alternatively used, generating a tissue-specific and stimulus-induced pattern of BDNF expression.[32,33] In particular, we reported that wild-type huntingtin target within the BDNF promoter II is a silencer sequence called RE1/NRSE (repressor element-1/neuronal restrictive silencer element) and its binding protein REST/NRSF (repressor element-1 transcription factor/neuron restrictive silencer factor). [31]NRSE like-elements have been found in several genes that have a key role for the terminal differentiation and correct functioning of neurons, including ion channels, neurotrophins, neurotransmitter biosynthetic enzymes, synaptic vesicle components and neuronal cell adhesion molecules. [34] Recently, our group has found that wild-type huntingtin binds and sequesters REST/NRSF in the cytoplasm, thereby preventing the formation of the nuclear co-repressor complex and allowing the expression of NRSE-controlled genes.[31] Dysfunction in huntingtin due to the presence of the CAG expansion makes huntingtin less able to bind to REST/NRSF, leading to its aberrant accumulation in the nucleus and to BDNF gene repression as well as to repression of a subset of other neuronal NRSE-controlled genes. This finding opens the possibility of using the NRSE as a target for the development of molecules able to inactivate its silencing activity and that can be administered as a complementary approach to cell replacement (FIG. 4).

In conclusion, stem cell therapy will be important, but probably not sufficient on its own, to treat neurodegenerative diseases. Alternative or complementary strategies should continue to be developed, which may include neuroprotective therapies, neurosurgical approaches, and cognitive and physical rehabilitation.[1]

ACKNOWLEDGMENTS

Our work is supported by Fondazione Telethon Onlus (Grant No. GGP02457) and Fondazione Cariplo (to L.C.) and by Fondo di Investimenti per la Ricerca di Base (No. RBNE01YRA3-001), Istituto Superiore di Sanità—Programma Nazionale sulle Cellule Staminali, CS117 (Italy), Ministero della Sanità—Ricerca Finalizzata Alzheimer 2000 (Italy) and the European Commission Framework VI Programme EUROSTEMCELL (to E.C.).

REFERENCES

1. ROSSI, F. & E. CATTANEO. 2002. Opinion: neural stem cell therapy for neurological diseases: dreams and reality. Nat. Rev. Neurosci. **3:** 401–409.
2. CONTI, L., T. CATAUDELLA & E. CATTANEO. 2003. Neural stem cells: a pharmacological tool for brain diseases? Pharm. Res. **47(4):** 289–297.
3. TEMPLE, S. 2001. The development of neural stem cells. Nature **414:** 112–117.
4. RAO, M.S. 1999. Multipotent and restricted precursors in the central nervous system. Anat. Rec. **257:** 137–148.
5. REYNOLDS, B.A. & S. WEISS. 1992. Generation of neurons and astrocytes from isolated cells of the adult mammalian central nervous system. Science **255:** 1707–1710.
6. VESCOVI, A.L., B.A. REYNOLDS, D.D. FRASER, *et al.* 1993. bFGF regulates the proliferative fate of unipotent (neuronal) and bipotent (neuronal/astroglial) EGF-generated CNS progenitor cells. Neuron **11:** 951–966.
7. REPRESA, A., T. SHIMAZAKI, M. SIMMONDS, *et al.* 2001. EGF-responsive neural stem cells are a transient population in the developing mouse spinal cord. Eur. J. Neurosi. **14:** 452–462.
8. KALYANI, A., K. HOBSON & M.S. RAO. 1997. Neuroepithelial stem cells from the embryonic spinal cord: isolation, characterization, and clonal analysis. Dev. Biol. **186:** 202–223.
9. McKAY, R. 1997. Stem cells in the central nervous system. Science **276:** 66–71.
10. SUSLOV, O.N., V.G. KUKEKOV, T.N. IGNATOVA, *et al.* 2002. Neural stem cell heterogeneity demonstrated by molecular phenotyping of clonal neurospheres. Proc. Natl. Acad. Sci. USA **99:** 14506–14511.
11. SMITH, A.G. 2001. Embryo-derived stem cells: of mice and men. Annu. Rev. Cell Dev. Biol. **17:** 435–462.
12. CATTANEO, E. & PG. PELICCI. 1998. Emerging roles for SH2/PTB-containing adaptor proteins in the developing mammalian brain. Trends Neurosci. **21:** 476–481.
13. CONTI, L., S. SIPIONE, D. RIGAMONTI, *et al.* 2001. The ShcC adaptor signals for survival and neuronal differentiation of CNS progenitor cells. Nat. Neurosci. **4:** 587–596.
14. LAI, KM. & T. PAWSON. 2000. The ShcA phosphotyrosine docking protein sensitizes cardiovascular signaling in the mouse embryo. Genes Dev. **14:** 1132–1145.
15. ZHANG, L., V. CAMERINI, T.P. BENDER, *et al.* 2002. A non redundant role for the adapter protein Shc in thymic T cell development. Nat. Immunol. **3:** 749–755.
16. MIGLIACCIO, E., M. GIORGIO, G. MELE, *et al.* 1999. The p66(shc) adaptor protein controls oxidative stress response and life span in mammals. Nature **402:** 309–313.
17. CONTI, L., C. DE FRAJA, M. GULISANO, *et al.* 1997. Expression and activation of SH2/PTB-containing ShcA adaptor protein reflects the pattern of neurogenesis in the mammalian brain. Proc. Natl. Acad. Sci. USA **94:** 8185–8190.
18. PONTI, G., L. CONTI, T. CATAUDELLA, *et al.* 2005. Comparative expression profiles of ShcB and ShcC phosphotyrosine adapter molecules in the adult brain. Neuroscience. In press.
19. O'BRYAN, J.P., Z. SONGYANG, L. CANTLEY, *et al.* 1996. A mammalian adaptor protein with conserved Src homology 2 and phosphotyrosine-binding domains is related to Shc and is specifically expressed in the brain. Proc. Natl. Acad. Sci. USA **93:** 2729–2734.
20. PELICCI, G., L. DENTE, A. DE GIUSEPPE, *et al.* 1996. A family of Shc related proteins with conserved PTB, CH1 and SH2 regions. Oncogen. **13:** 633–641.

21. NAKAMURA, T., S. MURAOKA, G. SANOKAWA, et al. 1998. N-Shc and Sck, two neu-
    ronally expressed Shc adapter homologs. J. Biol. Chem. **273:** 6960–6967.
22. PELICCI, G., F. TROGLIO, A. BODINI, et al. 2002. The neuron-specific Rai (ShcC) adap-
    tor protein inhibits apoptosis by coupling Ret to the phosphoatidylinosiyol 3-kinase/
    AKT pathway. Mol. Cell. Biol. **22:** 7351–7363
23. SAKAI, R., J.T. HENDERSON, J.P. O'BRYAN, et al. 2000. The mammalian ShcB and ShcC
    phosphotyrosine docking proteins function in the maturation of sensory and sympa-
    tetic neurons. Neuron **28:** 819–833
24. CATAUDELLA, T., L. CONTI & E. CATTANEO. 2003. Neural stem and progenitor cells:
    choosing the right Shc. Prog. Brain Res. **146:** 127–133.
25. CATTANEO, E., D. RIGAMONTI, D. GOFFREDO, et al. 2001. Loss of normal huntingtin
    function: new developments in Huntington's disease research. Trends Neurosci. **24:**
    182–188.
26. BACHOUD-LEVI, A.C., P. REMY, J.P. NGUYEN, et al. 2000. Motor and cognitive improve-
    ments in patients with Huntington's disease after neural transplantation. Lancet **356:**
    1975–1979.
27. GREENAMYRE, J.T. & I. SHOULSON. 2002. We need something better, and we need it
    now: fetal striatal transplantation in Huntington's disease? Neurology **58(5):** 675–
    676.
28. PESCHANKI, M. & S.B. DUNNETT. 2002. Cell therapy for Huntington's disease, the next
    step forward. Lancet Neurol. **1(2):** 81.
29. ZUCCATO, C., A. CIAMMOLA, D. RIGAMONTI, et al. 2001. Loss of Huntingtin-mediated
    BDNF gene transcription in Huntington's disease. Science. **293:** 493–498
30. BAQUET, Z.C., J.A. GORSKI & K.R. JONES. 2004 Early striatal dendrite deficits fol-
    lowed by neuron loss with advanced age in the absence of anterograde cortical brain-
    derived neurotrophic factor. J. Neurosci. **24:** 4250–4258.
31. ZUCCATO, C., M. TARTARI, A. CROTTI, et al. 2003. Huntingtin interacts with REST/
    NRSF to modulate the transcription of REST-controlled in neuronal genes. Nat.
    Genet. **35:** 76–83.
32. TIMMUSK, T., K. PALM, M. METSIS, et al. 1993. Multiple promoters direct tissue-spe-
    cific expression of the rat BDNF gene. Neuron **10:** 475–489.
33. TIMMUSK, T., U. LENDAHL, H. FUNAKOSHI, et al. 1995. Identification of brain-derived
    neurotrophic factor promoter regions mediating tissue-specific, axotomy-, and neu-
    ronal activity-induced expression in transgenic mice. J. Cell Biol. **128:** 185–199.
34. BRUCE, W.A., I.J. DONALDSON. I.C. WOOD, et al. 2004. Genome-wide analyses of
    repressor element 1 silencing transcription factors/neuron restrictive silencing factor
    (REST/NRSF) target genes. Proc. Natl. Acad Sci. USA **101:** 10458–10463.

Engineering a Dopaminergic Phenotype in Stem/Precursor Cells

Role of Nurr1, Glia-Derived Signals, and Wnts

ERNEST ARENAS

Laboratory of Molecular Neurobiology, Department of Medical Biochemistry and Biophysics, Karolinska Institute, 17 177 Stockholm, Sweden

ABSTRACT: Recent results from clinical trials using fetal tissue grafts in patients with Parkinson's disease (PD) have indicated that current surgical strategies for dopamine cell replacement therapy need to be improved in order to achieve better functional integration of the grafts and to avoid dyskinesias. Previous studies using rich dopaminergic (DA) cell suspensions have provided proof-of-concept that PD patients can benefit from cell replacement therapy. Stem cells have been proposed as better candidates for cell replacement therapy in PD since they can be standardized, expanded, and engineered *in vitro*. Recent developments indicate that cell preparations enriched in DA neurons can be generated *in vitro,* but their functional integration in animal models of disease is still far from optimal. This is not entirely surprising considering our limited knowledge of the development of DA neurons and the reduced number of factors that have been implemented in stem cell differentiation protocols. This review will focus on three aspects of DA neuron development: (1) the function of Nurr1 and retinoid X receptors (RXR) in the differentiation of DA precursors and in the survival of DA neurons; (2) the role of glia in DA neurogenesis and the differentiation of DA precursors; and (3) the function of the Wnt family of lipoproteins in the proliferation and differentiation of DA precursors. A greater understanding of the cellular and molecular mechanisms that control DA neuron development, as well as their functional integration *in vivo*, are likely to ultimately contribute to the development of novel stem cell replacement therapies for PD.

KEYWORDS: stem cells; dopamine; development; Parkinson's disease; neurodegeneration

ABBREVIATIONS: DA, dopaminergic; ESCs, embryonic stem cells: FCS, fetal calf serum; GDNF, glial cell line–derived neurotrophic factor; GFP, green fluorescent protein; LIF, leukemia inhibitory factor; NSCs, neural stem cells; PD, Parkinson's disease; RAR, retinoic acid receptor; RXR, retinoid X receptor; TH, tyrosine hydroxylase; VM, ventral midbrain (VM).

Address for correspondence: Ernest Arenas, Laboratory of Molecular Neurobiology, Department of Medical Biochemistry and Biophysics, Karolinska Institute, Scheeles vägen 1, 17 177, Stockholm, Sweden.

Ernest.Arenas@mbb.ki.se

Ann. N.Y. Acad. Sci. 1049: 51–66 (2005). © 2005 New York Academy of Sciences.
doi: 10.1196/annals.1334.007

CELL REPLACEMENT AND PARKINSON'S DISEASE

Parkinson's disease (PD) is a neurodegenerative disorder characterized by a triad of motor symptoms—tremor, rigidity, and slow movements (hypokinesia)—that have been associated with a predominant loss of substantia nigra dopaminergic (DA) neurons and a decrease in the levels of dopamine in the caudate and putamen. Current treatment is mainly based on the oral administration of L-DOPA, a precursor of the neurotransmitter that is taken up by DA neurons and is used to synthesize dopamine. The increased synthesis, storage, and release of dopamine in surviving DA nerve endings serves to compensate for the decreased levels of neurotransmitter in the caudate and putamen. However, as degeneration progresses, the treatment loses effectiveness in symptomatic relief, and adverse effects, including dyskinesias, occur. These obstacles underscore the need for alternative therapeutic tools to treat PD. Several approaches, including improved DA drugs, neuroprotective therapies, and cell replacement are currently being developed. Cell replacement is a candidate therapy with promising therapeutic potential, as demonstrated in 6-hydroxydopamine and MPTP models of Parkinson's disease.[1] Initial open-label clinical trials using human fetal ventral midbrain as material for transplantation have indicated that cell replacement can ameliorate not only [^{18}F]-fluorodopa uptake, but also the symptoms caused by the disease, permitting a reduction in the dose of L-DOPA required by the patients.[2,3] However, two recent double-blind clinical trials[4,5] described limited beneficial effects of fetal tissue grafts compared to placebo (sham surgery), and they induced off-phase dyskinesias. Several reasons could account for the low efficacy of the grafting procedure in the double-blind studies, including the low number of DA neurons in the graft, use of extruded tissue or tissue blocks and inappropriate immunosuppression.[6] Interestingly, subsequent analysis of the data from the double-blind trials[7,8] revealed that patients with good preoperative response to L-DOPA, younger than 60 years old, or grafted with four donors, experienced a functional recovery similar to that described in the open-label trials. These results suggest that grafts rich in DA neurons are necessary and that younger patients responsive to L-DOPA are preferred candidates for cell replacement therapy. The cause for the graft-induced dyskinesias it is still a matter of debate since they were more severe in patients transplanted with non-dissociated and pre-cultivated tissue,[4] two conditions that reduce the yield of surviving DA neurons, as compared to fresh cell suspensions. The availability of pure DA cell preparations, such as those obtained from TH-EGFP mice, will allow experimental examination of not only the therapeutic potential of DA neurons *per se*,[9,10] but also the relative contribution of DA and non-DA cells in ventral midbrain (VM) graft-induced dyskinesias. Additional factors that also remain to be examined include the inflammatory or immune reaction, the site of transplantation, the number of cells per site, and their migration. These factors may lead to a patchy distribution of cells, resulting in an irregular innervation of target structures, and an uneven DA neurotransmission that could favor the appearance of dyskinesias. As discussed above, the discrepancies between open-label and double-blind clinical trials, have raised a number of important issues that need to be addressed to develop a DA cell replacement therapy suitable for PD. A common obstacle has been the difficulty in obtaining sufficient human VM fetal tissue and/or DA neurons with standard quality and good viability. All of these issues have required researchers to examine alternative sources of DA cells for transplantation, such as stem cells.

NEURAL STEM CELLS AS CANDIDATES FOR DA CELL REPLACEMENT

Stem cells are cells with capacity to self-renew and differentiate into multiple cell types, including all lineages in the nervous system (neural stem cells, NSCs) or all cells in an individual (embryonic stem cells, ESCs). These properties, which allow indefinite expansion of the cells *in vitro* and their integration in the adult brain, have placed stem cells as optimal candidates for cell replacement. Based on their capacity to differentiate into midbrain DA neurons, several types of stem cells have been proposed as putative candidates for cell replacement in PD. These include ESCs, fetal and adult NSCs, and non-neural adult stem/progenitor cells from the bone marrow.[11–13] Most of the work in this area initially focused on the differentiation of NSCs or neural precursors. DA neurons have been obtained from a broad range of cell types, methods, and culture conditions, including:

(1) a v-myc immortalized mouse P4 cerebellar neural stem cell line overexpressing Nurr1 (Nurr1-c17.2) that differentiated by 80% into DA neurons in the presence of VM glia and RXR ligands *in vitro*[14];

(2) E12 rat VM DA precursors expanded in bFGF that differentiated by 41% into DA neurons (56% of the neurons) in ascorbic acid and low oxygen *in vitro*[15–17];

(3) E14 primary cultures from mouse striatum, cortex, or subependyma, expanded in bFGF that differentiated up to 5.5% into TH$^+$ and GABA$^+$ cells in serum or conditioned media from the B49 cell line[18];

(4) E14.5 rat mesencephalic progenitors expanded in EGF that differentiated into DA neurons by 80% in fetal calf serum (FCS, 10%) and a cytokine cocktail (interleukins-1 and -11, leukemia inhibitory factor [LIF], and glial cell line–derived neurotrophic factor [GDNF])[19];

(5) week 5–11 human midbrain precursors expanded in bFGF and EGF, at low O_2, that differentiated by 1% into DA neurons on a monolayer of striatal cultures with the above cytokine cocktail and 10% FCS[20];

(6) a v-myc immortalized human NSC line (hNS1) or neurospheres overexpressing Bcl-X(L) expanded in bFGF + EGF that differentiated by 5–8% into TH$^+$ cells with TPA and aFGF.[21]

The yield of DA neurons from neural stem/precursor cells *in vitro* has been variable, and as shown above, ranges from about 5% in the human hNS1 cell line[21] or mouse striatal precursors[18] to about 80% in rodent neural stem or precursor cell lines.[14,19] However, the main difficulty has been to obtain cells that survive the grafting procedure in animal models of PD. Recently, Liste *et al.*[21] found that human DA neurons survive after grafting in animal models of PD, but in low numbers. These results are important because they provide evidence that human NSCs are indeed candidate therapeutic tools in PD. However, the most successful methods in terms of achieving surviving TH$^+$ cells after grafting gave rise to only 0.13%[19] or 0.35%[16] TH$^+$ cells out of the total cells that were grafted. These numbers are still very low and reflect the fact that protocols for the effective integration of DA-derived NSCs *in vivo* are still far from optimal. Initially, NSCs because of their neural origin were expected to more easily be able to differentiate and give rise to any neuronal cell type. However, one problem of cultured NSCs is that they change their repertoire of transcription factors when propagated *in vitro* and certain neural phenotypes are difficult to obtain. NSCs have been reported to change the expression levels of region-specific transcription in response to culture conditions.[22,23] This change in transcription factor profile likely accounts for at least some of the difficulties in differentiating NSCs into specific cell types, since their responsiveness to signals delivered *in vitro* would not necessarily be the same as *in vivo*. In addition to the transcription factor(s) initially expressed by the NSCs, suboptimal basic cultivation protocols may contribute to the difficulties in the differentiation, and differences in the protocols contribute to the variability of the responses to differentiation-inducing signals. The solution might involve brief *in vitro* cultivation periods and optimized culture conditions using region-specific factors to mimic the conditions that the NSCs encounter *in vivo*. This may allow NSCs to retain or re-program the expression of transcription factors and activate signaling pathways required for the adequate acquisition of specific phenotypes and the subsequent integration of the cells after grafting *in vivo*.

EMBRYONIC STEM CELLS AS CANDIDATES FOR DA CELL REPLACEMENT

The differentiation of ESCs into DA neurons is the alternative that currently seems more promising. First, the repertoire of transcription factors that hESCs express is remarkably similar,[24,25] making ESCs a well standardized and homogenous cell preparation. Second, ESCs respond to defined differentiation signals in a highly reproducible manner and the efficiency at which they differentiate into DA neurons is quite high. Mouse ES cells have been differentiated into DA neurons following two basic protocols, based either on a five-step differentiation to neural and DA cells[26] or by unknown factors delivered by bone marrow stromal cell lines in co-culture.[27,28] Complementary strategies aiming at selecting and enriching DA neurons *in vitro*, prior to transplantation, have been developed. These include the expression of a selectable marker (green fluorescent protein, GFP) under the TH[29] or Pitx3[31] promoter to allow subsequent cell sorting and the transplantation of DA-rich cell preparations.[30] An alternative to this strategy, also based on GFP, may be the selection of committed DA precursors, using promoters of genes expressed in DA precursors such as Nurr1, ADH-2 or perhaps neurogenic genes. This approach would allow the differentiation of the precursors in the host brain and may contribute to reduce the loss of DA neurons during transplantation. The identification of endogenous proteins selectively expressed at the plasma membrane of DA precursors/neurons would also allow the isolation of cells by FACS sorting without the need of genetic modification. Other strategies have aimed at improving the DA differentiation of mESCs by overexpressing transgenes such as the orphan nuclear receptor Nurr1,[32,33] which was previously found to improve the differentiation of NSCs.[14] Overexpression of BclxL, an antiapototic gene that enhances the survival and permits the differentiation of ES cells, has also proven to increase the yields of DA neurons from mESCs.[34] However, challenges have arisen when the safety, survival, and functional integration of the ESC-derived DA neurons have been evaluated *in vivo*. The main issue with regard to the safety of ESC preparations is that grafting of undifferentiated cells can give rise to teratocarcinomas.[28,35] It is generally believed that if adequate differentiation or selection protocols are developed for ESCs, the risk for tumor formation might be very significantly reduced or even eliminated. Research in this area has focused on developing protocols that either eliminate or reduce the risk for tumor formation, even when undifferentiated ESCs are transplanted. Initial focus has been placed on the deletion of genes that are not required for DA neuron development but that are involved in tumor formation. One such candidate gene is cripto, a GPI-anchored EGF-CFC receptor that promotes cardiomyocyte differentiation while preventing the neural differentiation of ES cells.[36,37] Moreover, cripto is overexpressed in epithelial cancers, and blocking antibodies against cripto have been shown to suppress tumor cell growth.[38–40] Interestingly we found that cripto[−/−] mESCs readily differentiate into DA neurons upon treatment with Shh and FGF8 *in vitro*.[41] Moreover, grafting of 2000 undifferentiated cripto[−/−] mES cells in the adult striatum of animals receiving ipsilateral 6-OHDA resulted in enhanced behavioral recovery after seven weeks and no tumor formation was detected.[41] Instead, cripto[+/+]cells gave rise to tumors in 80% of the grafted animals. Thus, our results indicate that the safety and DA differentiation of ES cells can be enhanced by the deletion of cripto. These data suggest that the deletion of undesired genes, such as cripto, may constitute a novel strategy to enhance the safety and therapeutic potential of ES cells in cell replacement therapies for neurological diseases, including PD.

Another difficulty in the field has been that protocols for the differentiation of mESCs, when applied to hESCs, result in human DA neurons *in vitro* that do not survive well *in vivo*. It has been reported that the co-cultivation of hESCs with bone marrow stromal cell lines can be applied successfully to the differentiation of non-human primate ESCs[42] and human ESCs[43,44] *in vitro*. However, the differentiation of hESCs into DA neurons capable of surviving and functionally replacing DA cell loss in animal models of disease *in vivo* remains a challenge. Only a recent report[45] has showed that hES-derived DA neurons survive after transplantation, albeit at low numbers, in a rat 6-OHDA lesion model of PD. The differentiation protocol in that study is based on the administration of conditioned media from the Hep G2 cell line and bFGF. This protocol gave rise to functional DA neurons *in vitro* but their function *in vivo* was not assessed owing to the low number of surviving cells observed after grafting.[45] This is a crucial point in the development of stem cell therapies since human ventral mesencephalic tissue remains the gold standard in cell replacement. To date, the survival and/or the degree of differentiation of hESC-derived DA neurons after transplantation are not as good as those observed in grafts of human ventral mesencephalic cells in suspension. These observations clearly underscore the need to unravel the molecular mechanisms that control DA differentiation and survival in more detail.

Most of the current protocols for ESC differentiation discussed above are based on very early developmental signals. The first genes important for the formation of ventral midbrain DA neurons during development are those that provide regional identity to the neural tube (patterning) by establishing both anteroposterior and dorsoventral patterns of gene expression. Subsequently, from embryonic day (E) 9 to E10.5, there are no newly expressed genes that have been identified as participating in DA neuron development. During this period, DA precursor cells complete their specification and differentiate into DA neurons by E11.5. Two key players in midbrain patterning are sonic hedgehog (Shh), which provides ventral identity, and fibroblast growth factor 8 (FGF8), a key component of the midbrain-hindbrain organizer (for review see Prakash and Wurst[46]). Shh and FGF-8 have been successfully used to differentiate embryonic stem cells (ESCs) into DA phenotypes using

either the five-step differentiation protocol[26] or in co-culture with bone marrow stromal cell lines.[28,43,44] Many of these protocols give rise to DA neurons with adequate
electrophysiological properties, but as it may be expected from signals that provide
regional identity (Shh and FGF8), all give rise to non-DA midbrain and hindbrain
cell types, including glial cells and neurons with GABAergic, serotonergic, and cho-
linergic phenotypes.[26,28,32,43] In the most successful protocols, 60% DA neurons
and 40% non-DA cells were generated, transplanted, and integrated in the host brain.
The integration of non-DA neurons in the brain circuitry is likely to increase the levels of non-DA neurotransmission and may create aberrant or unnecessary connections that may hamper functional recovery and exacerbate or cause adverse effects.
It is therefore important to minimize the number of non-DA neurons that are transplanted. This might be achieved by selecting DA neurons or DA precursors prior to
grafting with cell sorting, using reporter genes (GFP) with specific promoters,[29–31]
or specific antibodies against membrane epitopes. A second possibility would be to
selectively instruct a DA phenotype. However, this possibility is complicated by the
fact that there are many unknown factors required for the development of midbrain
DA neurons. Since some of these unknown factors derive from glial cells present in
the vicinity of the newborn DA neurons,[14,47] one possible future replacement therapy could involve the transplantation of both DA neurons and the appropriate glial
cells that normally accompany DA neurons, to reconstruct the microenvironment in
which DA neurons develop. This might be particularly important in the case of human ESCs since protocols similar to those developed for mouse ESC have not been
successful in producing human DA neurons that engraft and contribute to a functional or behavioral recovery in PD models. It remains to be determined whether species
differences in the developmental pathways and/or the timing of ESC development
compromises the induction, differentiation or survival of human DA neurons from
ESCs. Interestingly, human development takes a longer time than mouse development, and mature glial cells are generated much later after the birth of DA neurons
in humans than in rodents. It is therefore possible that the lack of sufficient glial cells
and glial-derived signals in hESC preparations might in part contribute to the lower
survival efficiency of hESC-derived DA neurons after grafting. Indeed, it is becoming increasingly evident that additional factors that either interfere with and/or are
required for the proper function, maintenance and/or survival of DA neurons, have
not yet been accounted for. Surprisingly, our knowledge of the cascades triggered by
patterning factors or the signals involved in neurogenesis or prenatal differentiation
of DA neurons are largely unknown. A deeper knowledge of the mechanisms that
control the development of DA neurons is likely to provide us with additional target
genes and tools required to: (1) implement better and more complete DA differentiation protocols; (2) improve the functional integration of stem cell-derived DA neurons *in vivo*; and (3) contribute to the development of better stem cell replacement
therapies for the treatment of PD. Some of these factors are discussed below.

DIFFERENTIATION OF DA PRECURSORS/STEM CELLS: ROLE OF NURR1 AND GLIA-DERIVED SIGNALS

Nurr-1 is an orphan nuclear receptor that is highly expressed in DA precursors
and DA neurons in the VM. Nurr1 is a transcription factor that belongs to the thyroid

hormone/retinoic acid nuclear receptor superfamily. The three-dimensional structure of Nurr-1 has been recently resolved, and unlike other members of this family, the pocket for ligand binding is filled by hydrophobic residues, suggesting that Nurr1 is indeed a true orphan nuclear receptor.[48] With regard to the function of Nurr1 in brain development, deletion of the Nurr1 gene results in the loss of ventral midbrain DA neurons by their time of birth, at embryonic day 11.5.[49–51] Of interest, DA neurons in other brain regions are born and develop normally, indicating that ventral midbrain DA neurons are specifically lost. These reports demonstrated that Nurr1 is required for the birth of DA neurons and led many investigators to examine whether Nurr1 was also sufficient to induce a DA phenotype. Overexpression of Nurr1 in a NSC cell line (c17.2), which does not spontaneously generate TH$^+$ cells, resulted in cells that upon removal of mitogens predominantly differentiated into neurons, but less than 1% were TH-positive.[14] These results indicated that Nurr1 was not sufficient *per se* to promote DA differentiation and suggested that additional factors could be involved. Several known factors, including Shh and FGF8, were examined, but no effect on DA differentiation of Nurr1-NSCs was detected. To assess whether novel factors or factors with unknown function on DA precursors/neurons could be involved, co-cultures through a microporous insert with glial cells isolated from different brain regions were performed. Interestingly, co-culture with ventral midbrain glia promoted DA differentiation of Nurr1-transfected NSCs.[14] This observed effect was very robust, given that 80–90% of the neurons adopted a DA phenotype and 80% of the neural stem cells differentiated into neurons. Of interest it was noted that glial cells from other brain regions, such as the cortex, hippocampus or spinal cord, did not promote DA differentiation of Nurr1-NSCs in co-culture, but promoted their differentiation into morphologically distinct neuronal cell types. These results suggested a model in which: (1) Nurr1 in combination with cell autonomous neurogenic signals promotes neuronal differentiation of stem cells; and (2) that poorly soluble non-autonomous signals derived from glial cells contribute to promote the differentiation of Nurr1-positive stem or precursor cells into specific neuronal phenotypes. Provided that neurogenesis precedes gliogenesis during development and the major glial cell type present in the brain before neurogenesis is the radial glia, we have previously proposed[47] that radial glia might be the cell type providing such signals. In agreement with this possibility, glial cell preparations enriched in astrocytes and in glia-derived DA-inducing activity also contained cells with radial glia morphology (see Figure 5B in Wagner *et al.*[14]). Moreover, since several reports have demonstrated that radial glia can give rise to neurons in different brain structures,[52–59] it is possible that such cells could be both the source of neurons and instructive/ differentiation signals for neuronal precursors. We therefore hypothesized that VM radial glia could divide asymmetrically to generate neuronal precursors that would then respond to radial glia-derived factors and differentiate into DA neurons (FIG. 1A). Alternatively, neural stem cells may give rise to both radial glia and neuronal precursors, allowing radial glia to provide soluble factors to the neuronal precursors (FIG. 1B). Lineage tracing analysis should distinguish between these two models. Finally, radial glial cells may terminally differentiate into astrocytes and keep both physical contact and exchange diffusible factors both during development and into adulthood. During early stages, the radial glia may produce inductive signals or signals promoting differentiation, and at later stages astrocytes may be more involved in promoting neuronal survival and in the maintenance of the neuronal phenotype.[47]

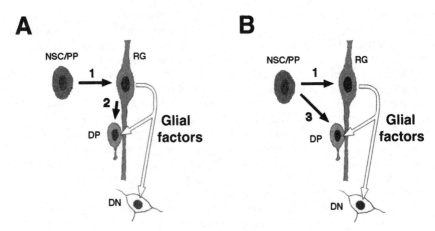

FIGURE 1. Alternative models on the possible lineage relations (*solid arrows*) of dopaminergic precursors (DPs) during development. In both models, neural stem cells (NSCs) or proliferative precursors (PPs) give rise to radial glia (RG, 1), and radial glia provide instructive, differentiation and survival factors (*open arrows*) to dopaminergic precursors and dopaminergic neurons (DN). In model **A**, radial glia give rise to dopaminergic precursors (2); while in model **B**, neural stem cells or proliferative precursors give rise to dopaminergic precursors (3).

Notably, the importance of helper cells such as VM glia in the DA differentiation of stem cells has only been realized in conditions where glial cells are not generated, such as Nurr1-c17.2 NSCs.[14] Indeed, other Nurr1-overexpressing NSCs[60,61] or ESCs[32,33,62] give rise to glial cells that could then inadvertently participate in the differentiation of DA precursors *in vitro*. Thus, some of the outstanding issues at this point are: (1) to identify glia-derived signals that promote the differentiation of neural stem/precursor cells into DA neurons,[14,47] and (2) to examine whether these factors are also part of the stromal derived DA inductive activity (SDIA) that have been reported to induce a DA phenotype in embryonic stem cells.[27,28,42–44] One candidate survival signal derived from glial cells could be docosahexaenoic acid (DHA), a ligand for the retinoid X receptor (RXR)–Nurr1 heterodimer that has recently been reported to promote the survival of newborn DA neurons.[63] Indeed, telencephalic radial glia-like cells have been reported to be the source of DHA,[64] but to date it remains unknown whether ventral midbrain radial glia are a source of DHA. Regardless of whether DHA is the ligand for the RXR-Nurr1 heterodimer in the VM, it is clear that endogenous ligands that work on the RXR-Nurr1 heterodimer exist *in vivo*. Indeed, reporting activity has been detected in the VM of dimerization-competent Nurr1-Gal4 transgenic mice, but not in dimerization-deficient Nurr1-Gal4 mice, indicating that RXR-Nurr1 dimerization is necessary for transactivation *in vivo*.[63] This activity was detected in a VM domain where DA neurons are born and where DA precursor cells express Aldh1a1, an enzyme involved in retinoid synthesis. Interestingly, both RXR agonists and RAR antagonists increased the survival of DA neurons *in vitro*, suggesting that the balance between RXR and RAR controls the survival of midbrain DA neurons[63] (FIG. 2). In agreement with a key role of RXR ligands in the development of DA neurons, we have previously reported that RXR

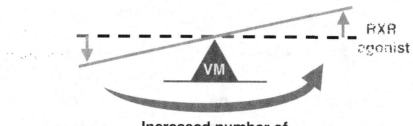

Increased number of dopaminergic neurons

FIGURE 2. The balance between retinoic acid receptor (RAR) and retinoid X receptor (RXR) signaling determines the number of dopaminergic neurons in the developing ventral midbrain (VM).[60] A balance in favor of RXR signaling, induced by either RXR agonists or RAR antagonists, increases the number of dopaminergic neurons. Instead, a balance in favor of RAR signaling decreases the number of dopaminergic neurons.

agonists can enhance the DA differentiation of Nurr1-overexpressing NSCs.[14] Thus, combined, our results suggest that retinoids play a key role in the development of midbrain DA neurons by possibly promoting the differentiation of DA precursors and enhancing the survival of DA neurons.

FUNCTION OF WNTS IN DOPAMINERGIC NEURON DEVELOPMENT

The Wnt family of lipoproteins is composed of 19 members. Wnts are known to bind and activate two different classes of receptors, the *frizzleds*, a family of 10 seven-pass transmembrane receptors, and two low-density lipoprotein-related receptor proteins (LRPs) receptors, LRP-5 and -6. Wnts have been shown to regulate stem cell self-renewal in hematopoietic cells,[65] proliferation,[66–70] fate decisions,[71–76] and differentiation in neural cells.[77–80] Wnt-1 is expressed by E8 in the midbrain/hindbrain boundary and by E10.5 its expression specifically extends to the ventral midbrain, to the domain where DA neurons are born. Interestingly, Wnt-1-null mutant mice exhibit a deletion of the anterior hindbrain and posterior midbrain, including the region where midbrain DA neurons are found.[81,82] These results indicate that Wnt-1 is required for the adequate patterning of this area and for the proliferation and/or differentiation of neural precursors. Interestingly, a similar phenotype has been reported for one of the Wnt co-receptors, LRP6.[83] Real-time RT-PCR analysis indicated that Wnt-1 and -5a are two of the Wnts expressed in the ventral midbrain at the highest levels the day that DA neurons are born. Subsequent *in situ* hybridization studies revealed that both Wnts are expressed by E10.5 in the mice in a Nurr1[+] domain that one day later gives rise to DA neurons. Moreover, Nurr1[+] cells were also found to express β-catenin, and to possibly signal through β-catenin by E10.5, as suggested by the detection of beta-Gal reporting activity in TOP-Gal mice in the VM.[84] To examine whether Wnt-1 and Wnt-5a indeed regulated the development of VM DA neurons, the effects of Wnt proteins were studied in VM precursor cultures. Studies on the function of Wnt proteins *in vitro* have been hampered by the poor sol-

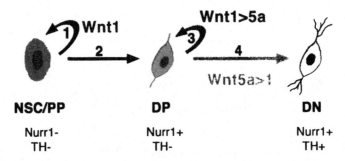

FIGURE 3. Model of the functions of Wnts in the development of dopaminergic neurons. Neural stem cells (NSCs) or proliferative precursors (PPs) in the midbrain region proliferate in response to Wnt1,[67] leading to the expansion of the precursor pool (1) without preventing their differentiation (2). Dopaminergic precursors proliferate *in vitro* in response to Wnt-1 to a greater extent than in response to Wnt-5a (3).[81] Instead, dopaminergic precursors differentiate into dopaminergic neurons to a greater extent by Wnt-5a than by Wnt-1 (4).[81,83]

ubility of Wnts, which is conferred by a palmytoyl moiety required for its biological activity.[85] However, in the presence of albumin, partially purified conditioned media from Wnt-overexpressing fibroblast cell lines were found to contain hemagglutinin-tagged Wnts.[84] Dose–response experiments with these media showed that Wnt-1 and -5a, but not Wnt-3a, increased the proportion of TH+ cells in E14.5 rat VM precursor cultures. Wnt-1 increased the overall number of TuJ1+ neurons in the VM, including TH+ neurons, by a mechanism involving the proliferation of VM precursors. However, Wnt-1 also permitted the differentiation of precursors into DA neurons, suggesting a sequential effect on proliferation and differentiation of DA precursors.[84] In agreement with this, transgenic animals overexpressing Wnt-1 under the control of the engrailed-1 promoter resulted in an increase in the number of cells and the overall size of all midbrain-hindbrain structures by a mechanism involving both proliferation and differentiation.[70] Instead, Wnt-5a enhanced the proliferation of DA precursors to a lesser extent and was found to regulate the differentiation of DA precursors into DA neurons and to induce the expression of c-ret mRNA,[84] the GDNF receptor required by newborn DA neurons for their survival (FIG. 3). Finally, Wnt-3a promoted BrdU incorporation and the proliferation of DA precursors, but did not allow their differentiation into DA neurons, resulting in fewer TH+ cells than in control cultures.[84] These effects were specific since they were not induced by control-conditioned media and the effect of Wnt-5a was partially blocked by a Wnt-blocking reagent, a fusion of the cysteine-rich domain of the Frizzled-8 receptor to the Fc fragment of human immunoglobulin. Subsequently, following the protocol published by Willert *et al.*[85] for the purification of Wnt-3a, we purified Wnt-5a and confirmed that Wnt-5a protein increases the number of TH+ neurons in E14.5 ventral midbrain precursor cultures in a dose-dependent manner.[86] Since Wnt-1 activates canonical Wnt signaling (mediated by β-catenin) and Wnt-5a activates non-canonical pathways, we examined whether the mechanism by which Wnt-5a increased the number of TH+ cells was by activating non-canonical signal-

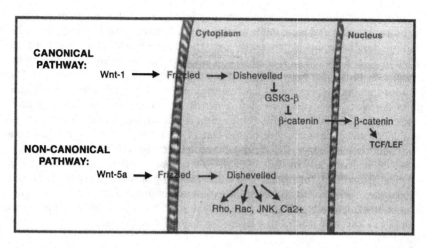

FIGURE 4. Simplified representation of the Wnt canonical and non-canonical pathways. Canonical Wnts, including Wnt1, bind to frizzled receptors and Lrp5/6 receptors to activate disheveled, which in turn inhibits glycogen synthetase kinase 3β (GSK3 β) and prevents the phosphorylation and degradation of β-catenin, which is stabilized in the cytoplasm and gains access to the nucleus, where it binds to TCF/LEF and regulates transcription. Noncanonical Wnts, including Wnt-5a, also bind to frizzleds and activates disheveleds, but this pathway leads instead to activation of the small GTPases, Rho and Rac, to the activation of jun kinase (JNK), and the activation of calcium pathways.

ing. Our results indicate that Wnt-5a increased Dishevelled phosphorylation but did not modify the levels of active, unphosphorylated, β-catenin in a midbrain DA cell line SN4741[87], suggesting that Wnt-5a activates non-canonical signaling in DA cells[86] (FIG. 4). However, overexpression of β-catenin or stabilization of β-catenin through inhibition of glycogen synthetase kinase-3 in E14.5 VM precursor cultures also increased the number of DA neurons through a mechanism involving the conversion of Nurr1 precursors into TH+ neurons.[88] These results suggest that the differentiation of DA precursors into TH+ neurons could be achieved by either signaling via β-catenin or via non-canonical pathways. It remains to be determined whether this is the case *in vivo* and whether both pathways may interact or cooperate to regulate DA differentiation in VM precursors.

FUTURE DIRECTIONS

Stem cell therapy continues to be an excellent candidate for regenerative medicine. However, this field is still in its infancy and much remains to be learned from both successful and unsuccessful approaches. The double-blind clinical trials with fetal tissue transplantation have not yielded the most positive results, but clearly, they have provided very valuable information concerning the inclusion criteria of patients for cell replacement therapy (young, L-DOPA-responsive patients) and cell preparations. The conclusions from these trials are therefore very important for the future design of clinical trials with stem cells for PD. For instance, cell preparations (sufficient DA cell numbers, minimizing non-DA neurons) and surgical approaches (multiple small deposits in denervated areas) need to be extensively studied and validated in animal models of disease. The trials have also underlined the need to understand the biological basis of the graft-induced dyskinesias, and to re-evaluate the need for immunosuppression at least during the first year after receiving allogenic grafts. Alternatively autologous adult stem cells or somatic nuclear transfer in ESCs or selection of immunocompatible donors from stem cell banks could also reduce the immunological/inflammatory response. One of the advantages of stem cells is that they could be selected *in vitro* for the desired phenotypic and functional properties and could be genetically modified *in vitro* to serve specific functions. Cells could be genetically modified, for instance, to either express genes necessary for DA development (i.e., Nurr1) or to eliminate the expression of genes that promote the generation of unwanted cells or tumors (i.e., cripto). In addition, stem cells could be used to engineer survival factors to prevent the loss of endogenous DA neurons (i.e., GDNF[89]).

It is also clear from the results from the clinical trials with fetal tissue grafts that before stem cells are utilized, we must achieve cell cultures that perform at least as safely and as well as fetal tissue grafts in terms of functional integration in animal models of PD. Focus needs to be placed in understanding the cellular and molecular mechanisms that control DA neuron development (Wnts and neurogenic factors) and maintenance after transplantation (survival and differentiation factors). One emerging theme of potential importance is the characterization of poorly soluble signals derived from radial glia, such as retinoids and novel, yet unidentified factors. Indeed glial cells seem to be the source of short-range lipophilic signaling molecules that guide the development of DA neurons and work to maintain their function. Similarly, Wnts have recently been found to be key regulators of DA neuronal development and could therefore contribute to develop novel stem cell differentiation protocols. These signals together with novel factors that are likely to be identified should have an important impact on the development of novel protocols for DA differentiation and the functional integration of DA neurons in the brain.

Finally, the ultimate goal of cell replacement therapy in Parkinson's disease should be to reconstruct all the brain circuitry affected by the disease. This implies that DA neurons would ideally need to be placed in the substantia nigra and their afferent and efferent connections re-established. Furthermore, non-DA neurons affected in PD should also be replaced. We should not be discouraged by the complexity of the problem, but we should be aware it. We are clearly still far from reaching these objectives, but there are many smaller objectives on the way to these goals that have the potential to change the course of the disease and/or offer symptomatic relief to patients suffering from Parkinson's disease.

ACKNOWLEDGMENTS

Dr. Carmen Salto and Kyle Sousa are acknowledged for critical reading of the manuscript, Lena Amaloo for secretarial help, and Claudia Tello for additional assistance. Financial support was obtained from the Swedish Foundation for Strategic Research, the Swedish Royal Academy of Sciences, the Knut and Alice Wallenberg

Foundation, the European Commission, the Michael J. Fox Foundation, the Juvenile Diabetes Research Foundation, the Swedish Research Council (V.R.), and the Karolinska Institute.

REFERENCES

1. HERMAN, J.P. & N.D. ABROUS. 1994. Dopaminergic neural grafts after fifteen years: results and perspectives. Prog. Neurobiol. **44:** 1–35
2. LINDVALL, O. & P. HAGELL. 2000. Clinical observations after neural transplantation in Parkinson's disease. Prog. Brain Res. **127:** 299–320
3. DUNNETT, S.B., A. BJÖRKLUND & O. LINDVALL. 2001. Cell therapy in Parkinson's disease: stop or go? Nat. Rev. Neurosci. **2:** 365–369.
4. FREED, C.R. *et al.* 2001. Transplantation of embryonic dopamine neurons for severe Parkinson's disease. N. Engl. J. Med. **344:** 710–719.
5. OLANOW, C.W. *et al.* 2003. A double-blind controlled trial of bilateral fetal nigral transplantation in Parkinson's disease. Ann. Neurol. **54:** 403–414.
6. BJORKLUND, A., S.B. DUNNETT, P. BRUNDIN, *et al.* 2003. Neural transplantation for the treatment of Parkinson's disease. Lancet Neurol. **2(7):** 437–445.
7. FREED, C.R., M.A. LEEHEY, M. ZAWADA, *et al.* 2003 Do patients with Parkinson's disease benefit from embryonic dopamine cell transplantation? J. Neurol. **250** (Suppl. 3): III44–46.
8. WINKLER, C., D. KIRIK & A. BJORKLUND. 2005. Cell transplantation in Parkinson's disease: how can we make it work? Trends Neurosci. **28:** 86–92.
9. SAWAMOTO, K., N. NAKAO, K. KOBAYASHI, *et al.* 2001. Visualization, direct isolation, and transplantation of midbrain dopaminergic neurons. Proc. Natl. Acad. Sci. USA **98:** 6423–6428.
10. YOSHIZAKI, T., M. INAJI, H. KOUIKE, *et al.* 2004. Isolation and transplantation of dopaminergic neurons generated from mouse embryonic stem cells. Neurosci. Lett. **363:** 33–37.
11. JIANG, Y., B.N. JAHAGIRDAR, R.L. REINHARDT, *et al.* 2002. Pluripotency of mesenchymal stem cells derived from adult marrow. Nature **418:** 41–49.
12. JIANG, Y., D. HENDERSON, M. BLACKSTAD, *et al.* 2003. Neuroectodermal differentiation from mouse multipotent adult progenitor cells. Proc. Natl. Acad. Sci. USA **100** (Suppl. 1): 11854–11860.
13. DEZAWA, M., H. KANNO, M. HOSHINO, *et al.* 2004. Specific induction of neuronal cells from bone marrow stromal cells and application for autologous transplantation. J. Clin. Invest. **113:** 1701–1710.
14. WAGNER, J., P. AKERUD, D.S. CASTRO, *et al.* 1999. Induction of a midbrain dopaminergic phenotype in Nurr1-overexpressing neural stem cells by type 1 astrocytes. Nat. Biotechnol. **17:** 653–659.
15. STUDER, L., V. TABAR & R.D.G. MCKAY. 1998. Transplantation of expanded mesencephalic precursors leads to recovery in parkinsonian rats. Nat. Neurosci. **1:** 290–295.
16. STUDER, L., M. CSETE, S.H., S.H. LEE, *et al.* 2000. Enhanced proliferation, survival, and dopaminergic differentiation of CNS precursors in lowered oxygen. J. Neurosci. **20:** 7377–7383.
17. YAN, J., L. STUDER & R.D. MCKAY. 2001. Ascorbic acid increases the yield of dopaminergic neurons derived from basic fibroblast growth factor expanded mesencephalic precursors. J. Neurochem. **76:** 307–311.
18. DAADI, M.M. & S. WEISS. 1999. Generation of tyrosine hydroxylase-producing neurons from precursors of the embryonic and adult forebrain. J. Neurosci. **19:** 4484–4497.
19. CARVEY, P.M., Z.D. LING, C.E. SORTWELL, *et al.* 2001. A clonal line of mesencephalic progenitor cells converted to dopamine neurons by hematopoietic cytokines: a source of cells for transplantation in Parkinson's disease. Exp. Neurol. **171:** 98–108.
20. STORCH, A., G. PAUL, M. CSETE, *et al.* 2001. Long-term proliferation and dopaminergic differentiation of human mesencephalic neural precursor cells. Exp. Neurol. **170:** 317–325.

21. LISTE, I., E. GARCIA-GARCIA & A. MARTINEZ-SERRANO. 2004. The generation of dopaminergic neurons by human neural stem cells is enhanced by Bcl-XL, both in vitro and in vivo. J. Neurosci. **24:** 10786–10795.

22. SANTA-OLALLA, J., J.M. BAIZABAL, M. FREGOSO, et al. 2003. The in vivo positional identity gene expression code is not preserved in neural stem cells grown in culture. Eur. J. Neurosci. **18:** 1073–1084.

23. SKOGH, C., M. PARMAR & K. CAMPBELL. 2003. The differentiation potential of precursor cells from the mouse lateral ganglionic eminence is restricted by in vitro expansion. Neuroscience **120:** 379–385.

24. SPERGER, J.M., X. CHEN, J.S. DRAPER, et al. 2003. Gene expression patterns in human embryonic stem cells and human pluripotent germ cell tumors. Proc. Natl. Acad. Sci. USA **100:** 13350–13355.

25. CARPENTER, M.K., E.S. ROSLER, G.J. FISK, et al. 2004. Properties of four human embryonic stem cell lines maintained in a feeder-free culture system. Dev. Dyn. **229:** 243–258.

26. LEE, S.H., N. LUMELSKY, L. STUDER, et al. 2000. Efficient generation of midbrain and hindbrain neurons from mouse embryonic stem cells. Nat. Biotechnol. **18:** 675–679.

27. KAWASAKI, H., K. MIZUSEKI, S. NISHIKAWA, et al. 2000. Induction of midbrain dopaminergic neurons from ES cells by stromal cell-derived inducing activity. Neuron **28:** 31–40.

28. BARBERI, T., P. KLIVENYI, N.Y. CALINGASAN, et al. 2003. Neural subtype specification of fertilization and nuclear transfer embryonic stem cells and application in parkinsonian mice. Nat. Biotechnol. **21:** 1200–1207.

29. SAWAMOTO, K., N. NAKAO, K. KOBAYASHI, et al. 2001. Visualization, direct isolation, and transplantation of midbrain dopaminergic neurons. Proc. Natl. Acad. Sci. USA **98:** 6423–6428.

30. YOSHIZAKI, T., M. INAJI, H. KOUIKE, et al. 2004. Isolation and transplantation of dopaminergic neurons generated from mouse embryonic stem cells. Neurosci. Lett. **363:** 33–37.

31. ZHAO, S., S. MAXWELL, A. JIMENEZ-BERISTAIN, et al. 2004. Generation of embryonic stem cells and transgenic mice expressing green fluorescence protein in midbrain dopaminergic neurons. Eur. J. Neurosci. **19:** 1133–1140.

32. KIM, J.H., J.M. AUERBACH, J.A. RODRIGUEZ-GOMEZ, et al. 2002. Dopamine neurons derived from embryonic stem cells function in an animal model of Parkinson's disease. Nature **418:** 50–56.

33. CHUNG, S., K.C. SONNTAG, T. ANDERSSON, et al. 2002. Genetic engineering of mouse embryonic stem cells by Nurr1 enhances differentiation and maturation into dopaminergic neurons. Eur. J. Neurosci. **16:** 1829–1838.

34. SHIM, J.W., H.C. KOH, M.Y. CHANG, et al. 2004. Enhanced in vitro midbrain dopamine neuron differentiation, dopaminergic function, neurite outgrowth, and 1-methyl-4-phenylpyridium resistance in mouse embryonic stem cells overexpressing Bcl-XL. J Neurosci. **24:** 843–852.

35. BJORKLUND, L.M., R. SANCHEZ-PERNAUTE, S. CHUNG, et al. 2002. Embryonic stem cells develop into functional dopaminergic neurons after transplantation in a Parkinson rat model. Proc. Natl. Acad. Sci. USA **99:** 2344–2349.

36. XU, C., G. LIGUORI, E.D. ADAMSON & M.G. PERSICO. 1998. Specific arrest of cardiogenesis in cultured embryonic stem cells lacking Cripto-1. Dev. Biol. **196:** 237–247.

37. PARISI, S., D. D'ANDREA, C.T. LAGO, et al. 2003. Nodal-dependent Cripto signaling promotes cardiomyogenesis and redirects the neural fate of embryonic stem cells. J. Cell Biol. **163:** 303–314.

38. PERSICO, M.G., G.L. LIGUORI, S. PARISI, et al. 2001. Cripto in tumors and embryo development. Biochim. Biophys. Acta **1552:** 87–93.

39. ADKINS, H.B., C. BIANCO, S.G. SCHIFFER, et al. 2003. Antibody blockade of the Cripto CFC domain suppresses tumor cell growth in vivo. J. Clin. Invest. **112:** 575–587.

40. SHEN, M.M. 2003. Decrypting the role of Cripto in tumorigenesis. J. Clin. Invest. **112:** 500–502.

41. PARISH, C.L., S. PARISI, M.G. PERSICO, et al. 2005. Cripto as a target for improving embryonic stem cell-based therapy in Parkinson's disease. Stem Cells **23:** 471–476.

42. KAWASAKI, H., H. SUEMORI, K. MIZUSEKI, et al. 2002. Generation of dopaminergic neurons and pigmented epithelia from primate ES cells by stromal cell-derived inducing activity. Proc. Natl. Acad. Sci. USA **99:** 1580–1585.

43. PERRIER, A.L., V. TABAR, T. BARBERI, *et al.* 2004. Derivation of midbrain dopamine neurons from human embryonic stem cells. Proc. Natl. Acad. Sci. USA **101**: 12543–12548.

44. ZENG, X., J. CHEN, J.F. SANCHEZ, *et al.* 2003. Stable expression of hrGFP by mouse embryonic stem cells: promoter activity in the undifferentiated state and during dopaminergic neural differentiation. Stem Cells **21**: 647–653.

45. SCHULZ, T.C., S.A. NOGGLE, G.M. PALMARINI, *et al.* 2004. Differentiation of human embryonic stem cells to dopaminergic neurons in serum-free suspension culture. Stem Cells **22**: 1218–1238.

46. PRAKASH, N. & W. WURST. 2004. Specification of midbrain territory. Cell Tissue Res. **318**: 5–14.

47. HALL, A.C., H. MIRA, J. WAGNER & E. ARENAS. 2003. Region-specific effects of glia on neuronal induction and differentiation with a focus on dopaminergic neurons. Glia **43**: 47–51.

48. WANG, Z., G. BENOIT, J. LIU, *et al.* 2003. Structure and function of Nurr1 identifies a class of ligand-independent nuclear receptors. Nature **423**: 555–560.

49. ZETTERSTROM, R.H., L. SOLOMIN, L. JANSSON, *et al.* 1997. Dopamine neuron agenesis in Nurr1-deficient mice. Science **276**: 248–250.

50. SAUCEDO-CARDENAS, O., J.D. QUINTANA-HAU, W.D. LE, *et al.* 1998. Nurr1 is essential for the induction of the dopaminergic phenotype and the survival of ventral mesencephalic late dopaminergic precursor neurons. Proc. Natl. Acad. Sci. USA **95**: 4013–4018.

51. CASTILLO, S.O., J.S. BAFFI, M. PALKOVITS, *et al.* 1998. Dopamine biosynthesis is selectively abolished in substantia nigra/ventral tegmental area but not in hypothalamic neurons in mice with targeted disruption of the Nurr1 gene. Mol. Cell. Neurosci. **11**: 36–46.

52. MALATESTA, P., E. HARTFUSS & M. GOTZ. 2000. Isolation of radial glial cells by fluorescent-activated cell sorting reveals a neuronal lineage. Development **127**: 5253–5263.

53. HARTFUSS, E., R. GALLI, N. HEINS & M. GOTZ. 2001. Characterization of CNS precursor subtypes and radial glia. Dev. Biol. **229**: 15–30.

54. MIYATA, T., A. KAWAGUCHI, H. OKANO & M. OGAWA. 2001. Asymmetric inheritance of radial glial fibers by cortical neurons. Neuron **31**: 727–741.

55. NOCTOR, S.C., A.C. FLINT, T.A. WEISSMAN, *et al.* 2002. Dividing precursor cells of the embryonic cortical ventricular zone have morphological and molecular characteristics of radial glia. J. Neurosci. **22**: 3161–3173.

56. SKOGH, C., C. ERIKSSON, M. KOKAIA, *et al.* 2001. Generation of regionally specified neurons in expanded glial cultures derived from the mouse and human lateral ganglionic eminence. Mol. Cell. Neurosci. **17**: 811–820.

57. HEINS, N., P. MALATESTA, F. CECCONI, *et al.* 2002. Glial cells generate neurons: the role of the transcription factor Pax6. Nat. Neurosci. **5**: 308–315.

58. GOTZ, M., E. HARTFUSS & P. MALATESTA. 2002. Radial glial cells as neuronal precursors: a new perspective on the correlation of morphology and lineage restriction in the developing cerebral cortex of mice. Brain Res. Bull. **57**: 777–788.

59. ANTHONY, T.E., C. KLEIN, G. FISHELL & N. HEINTZ. 2004. Radial glia serve as neuronal progenitors in all regions of the central nervous system. Neuron **41**: 881–890.

60. SAKURADA, K., M. OHSHIMA-SAKURADA, T.D. PALMER & F.H. GAGE. 1999. Nurr1, an orphan nuclear receptor, is a transcriptional activator of endogenous tyrosine hydroxylase in neural progenitor cells derived from the adult brain. Development **126**: 4017–4026.

61. KIM, J.Y., H.C. KOH, J.Y. LEE, *et al.* 2003. Dopaminergic neuronal differentiation from rat embryonic neural precursors by Nurr1 overexpression. J. Neurochem. **85**: 1443–1454.

62. SONNTAG, K.C., R. SIMANTOV, K.S. KIM & O. ISACSON. 2004. Temporally induced Nurr1 can induce a non-neuronal dopaminergic cell type in embryonic stem cell differentiation. Eur. J. Neurosci. **19**: 1141–1152.

63. WALLEN-MACKENZIE, A., A. MATA DE URQUIZA, S. PETERSSON, *et al.* 2003. Nurr1-RXR heterodimers mediate RXR ligand-induced signaling in neuronal cells. Genes Dev. **17**: 3036–3047

64. TORESSON, H., A. MATA DE URQUIZA, C. FAGERSTROM, *et al.* 1999. Retinoids are produced by glia in the lateral ganglionic eminence and regulate striatal neuron differentiation. Development **126**: 1317–1326.

65. REYA, T., A.W. DUNCAN, L. AILLES, *et al.* 2003. A role for Wnt signalling in self-renewal of haematopoietic stem cells. Nature **423**: 409–414.

66. TAIPALE, J. & P.A. BEACHY. 2001. The Hedgehog and Wnt signalling pathways in cancer. Nature **411:** 349–354.
67. CHENN, A. & C.A. WALSH. 2002. Regulation of cerebral cortical size by control of cell cycle exit in neural precursors. Science **297:** 365–369.
68. MEGASON, S.G. & A.P. MCMAHON. 2002. A mitogen gradient of dorsal midline Wnts organizes growth in the CNS. Development **129:** 2087–2098.
69. ZECHNER, D., Y. FUJITA, J. HULSKEN, et al. 2003. beta-Catenin signals regulate cell growth and the balance between progenitor cell expansion and differentiation in the nervous system. Dev. Biol. **258:** 406–418.
70. PANHUYSEN, M., D.M. VOGT WEISENHORN, V. BLANQUET, et al. 2004. Effects of Wnt1 signaling on proliferation in the developing mid-/hindbrain region. Mol. Cell Neurosci. **26:** 101–111.
71. DORSKY, R.I., R.T. MOON & D.W. RAIBLE. 1998. Control of neural crest cell fate by the Wnt signalling pathway. Nature **396:** 370–373.
72. BAKER, J.C., R.S. BEDDINGTON & R.M. HARLAND. 1999. Wnt signaling in *Xenopus* embryos inhibits bmp4 expression and activates neural development. Genes Dev. **13:** 3149–3159.
73. WILSON, S.I., A. RYDSTROM, T. TRIMBORN, et al. 2001. The status of Wnt signaling regulates neural and epidermal fates in the chick embryo. Nature **411:** 325–330
74. GARCIA-CASTRO, M..I, C. MARCELLE & M. BRONNER-FRASER. 2002. Ectodermal Wnt function as a neural crest inducer. Science **297:** 8488–8451.
75. MUROYAMA, Y., M. FUJIHARA, M. IKEYA, et al. 2002. Wnt signaling plays an essential role in neuronal specification of the dorsal spinal cord. Genes Dev. **16:** 548–553.
76. MUROYAMA, Y., H. KONDOH & S. TAKADA. 2004. Wnt proteins promote neuronal differentiation in neural stem cell culture. Biochem. Biophys. Res. Commun. **313:** 915–921.
77. HALL, A.C., F.R. LUCAS & P.C. SALINAS. 2000. Axonal remodeling and synaptic differentiation in the cerebellum is regulated by WNT-7a signaling. Cell **100:** 525–535.
78. KRYLOVA, O., J. HERREROS, K.E. CLEVERLEY, et al. 2002. WNT-3, expressed by motoneurons, regulates terminal arborization of neurotrophin-3-responsive spinal sensory neurons. Neuron **35:** 1043–1056.
79. YU, X. & R.C. MALENKA. 2003. Beta-catenin is critical for dendritic morphogenesis. Nat. Neurosci. **6:** 1169–1177.
80. ROSSO, S.B., D. SUSSMAN, A. WYNSHAW-BORIS & P.C. SALINAS. 2005. Wnt signaling through Dishevelled, Rac and JNK regulates dendritic development. Nat. Neurosci. **8:** 34–42.
81. MCMAHON, A.P. & A. BRADLEY. 1990. The Wnt-1 (int-1) proto-oncogene is required for development of a large region of the mouse brain. Cell **62:** 1073–1085.
82. THOMAS, K.R. & M.R. CAPECCHI. 1990. Targeted disruption of the murine int-1 proto-oncogene resulting in severe abnormalities in midbrain and cerebellar development. Nature **346:** 847–850
83. PINSON, K.I., J. BRENNAN, S. MONKLEY, et al. 2000. An LDL-receptor-related protein mediates Wnt signalling in mice. Nature **407:** 535–538.
84. CASTELO-BRANCO, G., J. WAGNER, F.J. RODRIGUEZ, et al. 2003. Differential regulation of midbrain dopaminergic neuron development by Wnt-1, Wnt-3a, and Wnt-5a. Proc. Natl. Acad. Sci. USA **100:** 12747–12752.
85. WILLERT, K., J.D. BROWN, E. DANENBERG, et al. 2003. Wnt proteins are lipid-modified and can act as stem cell growth factors. Nature **423:** 448–452.
86. SCHULTE, G., V. BRYJA, N. RAWAL, et al. 2005. Purified Wnt-5a increases dishevelled phosphorylation and differentiation of midbrain dopaminergic cells. J. Neurochem. **92:** 1550–1553.
87. SON, J.H., H.S. CHUN, T.H. JOH, et al. 1999. Neuroprotection and neuronal differentiation studies using substantia nigra dopaminergic cells derived from transgenic mouse embryos. J. Neurosci. **19:** 10–20.
88. CASTELO-BRANCO, G., N. RAWAL & E. ARENAS. 2004. GSK-3beta inhibition/beta-catenin stabilization in ventral midbrain precursors increases differentiation into dopamine neurons. J. Cell Sci. **117:** 5731–5737.
89. ÅKERUD, P., J. CANALS, E.Y. SNYDER & E. ARENAS. 2001. Neuroprotection through delivery of GDNF by neural stem cells in a mouse model of Parkinson's disease. J. Neurosci. **21:** 8108–8118.

Umbilical Cord Blood–Derived Stem Cells and Brain Repair

PAUL R. SANBERG,[a] ALISON E.WILLING,[a] SVITLANA GARBUZOVA-DAVIS,[a] SAMUEL SAPORTA,[b] GUOQING LIU,[a] CYNDY DAVIS SANBERG,[c] PAULA C. BICKFORD,[a] STEPHEN K. KLASKO,[a] AND NAGWA S. EL-BADRI[a]

[a]Center of Excellence for Aging and Brain Repair, Department of Neurosurgery, College of Medicine, University of South Florida, Tampa, Florida 33612, USA

[b]Department of Anatomy, College of Medicine, University of South Florida,, Tampa, Florida 33612.

[c]Saneron CCEL Therapeutics, Inc., Tampa Florida 33612, USA

ABSTRACT: Human umbilical cord blood (HUCB) is now considered a valuable source for stem cell–based therapies. HUCB cells are enriched for stem cells that have the potential to initiate and maintain tissue repair. This potential is especially attractive in neural diseases for which no current cure is available. Furthermore, HUCB cells are easily available and less immunogenic compared to other sources for stem cell therapy such as bone marrow. Accordingly, the number of cord blood transplants has doubled in the last year alone, especially in the pediatric population. The therapeutic potential of HUCB cells may be attributed to inherent ability of stem cell populations to replace damaged tissues. Alternatively, various cell types within the graft may promote neural repair by delivering neural protection and secretion of neurotrophic factors. In this review, we evaluate the preclinical studies in which HUCB was applied for treatment of neurodegenerative diseases and for traumatic and ischemic brain damage. We discuss how transplantation of HUCB cells affects these disorders and we present recent clinical studies with promising outcome.

KEYWORDS: stem cell; cord blood; neurogenesis; brain repair

INTRODUCTION

The last few years have witnessed an expansion in stem cell research and its potential for therapy following the revolutionary experiments in mammalian cloning. In addition to the controversial embryonic stem cell research, adult stem cell sources like hematopoietic stem cells, mesenchymal stem cells, epidermal stem cells, pancreatic stem cells, and several other organ stem cells are currently identified and characterized in laboratories all over the world. Clinical stem cell therapy dates back

Address for correspondence: Paul R. Sanberg, Ph.D., D.Sc., Distinguished Professor and Director, Center of Excellence for Aging and Brain Repair, Department of Neurosurgery, College of Medicine, University of South Florida, 12901 Bruce B. Downs Blvd., MDC 78, Tampa, FL 33612. Voice: 813-974-3154; fax: 813-974-3078.
psanberg@hsc.usf.edu

Ann. N.Y. Acad. Sci. 1049: 67–83 (2005). © 2005 New York Academy of Sciences.
doi: 10.1196/annals.1334.008

to the first bone marrow transplant experiments in the middle of the past century. Nonetheless, hematopoietic stem cell therapy (HSCT) has been reserved for life-threatening or advanced illness because of associated complications in the form of graft rejection, graft-versus-host disease (GVHD), and infections. These post-transplant bacterial and viral infections caused by conditioning regimens like chemotherapy, irradiation, cytokines, and the use of anti-T cell antibodies are considered the primary impediment for survival after stem cell transplantation.[1] Manipulation of the stem cell graft by enriching for desirable stem cells, deleting mature lymphocytes, and including additional stem cell sources like umbilical cord blood and peripheral blood stem cells has significantly improved the outcome of this form of stem cell therapy.[2,3]

Peripheral blood stem cells are rapidly becoming the standard care in autologous transplantation after the use of hematopoietic growth factors G-CSF and GM-CSF to recruit stem cells into peripheral blood. In the allogeneic setting, however, a major limitation to stem cell therapy has been the higher incidence of acute and chronic GVHD and its potential negative impact on survival.[4,5] During the past two decades, human umbilical cord blood (HUCB) has emerged as a novel valuable source for stem cells next to bone marrow and peripheral blood. Only 20 to 25% of patients are expected to find an HLA-matched sibling from the bone marrow donor pool. Since its first successful transplant for Fanconi's anemia in 1988, HUCB was found to be a highly enriched source for immature stem and blood cells that are less immunogenic than adult marrow and blood cells. Compared with bone marrow recipients, cord blood recipients from related or unrelated donors experience a decreased incidence of acute graft-versus-host disease and a rather delayed hematopoietic recovery.[6–8] Lower immunogenicity of cord blood is credited to abundant immature progenitors with longer telomeres than adult marrow stem cells.[9] T lymphocytes in the cord blood demonstrate a healthy proliferative response to alloantigen stimulation, but their cytotoxic lytic function seems to be depressed.[10,11] Cord blood–expanded dendritic cells (DCs) are less alloreactive in a mixed lymphocyte reaction than peripheral bloods DCs, which may account for the lower GVHD observed in cord blood. Decreased cytokine production like IL-12 and interferon gamma have been associated with lower immunogenicity.[13]

ENGRAFTMENT OF CORD BLOOD STEM CELLS

Unlike bone marrow transplants, limited data are available on HUCB engraftment and the dynamics of its contribution to reconstitution of the lymphohematopoietic system. Impressive evidence for engraftment was demonstrated when unrelated HUCB cells were transplanted into children with immune deficiency. Donor stem cells were engrafted, and induced rapid and reliable recovery of immune functions. In addition, these children suffered lower risk of GVHD and post-transplant infections.[14] In their elegant study, Traggiai et al.[15] injected CD34+ HUCB cells via the intrahepatic route into conditioned newborn immune-deficient Rag2$^{-/-}$gammac$^{-/-}$ mice. Cord blood cells engrafted and reconstituted primary and secondary lymphoid organs. De novo development of donor origin B and T lymphocytes, and dendritic cells was associated with production of normal functional immune responses.

Reconstitution studies show that short-term recovery of neutrophils is delayed compared to bone marrow and peripheral blood stem cells, but long-term recovery of T, B lymphocytes, and natural killer cells seem to be satisfactory.[16] The composition of cord blood cells may contribute to these dynamics. HUCB cells correlate with bone marrow cells with the exception of the lymphocyte content, which tends to be lowest in CB grafts.[17] The percentage of CD34+CD38− hematopoietic stem cells (HSCs) seems to be higher in cord blood as compared to bone marrow, as well as the natural killer (NK) cell populations.[18] Superior functions of cord blood stem cells have also been reported in hematopoietic colony functional assays. Pluripotent HSCs within cord blood had a higher cloning efficiency, were more proliferative in response to cytokine stimulation, and generated approximately seven-fold progeny when compared to bone marrow cells.[19] This enrichment was clinically supported by studies that showed the efficiency of fewer number of cord blood HSCs used in transplantation as compared to bone marrow.[20] Considering that one limitation of cord blood therapy in adults is the inadequate volume obtained from a single cord, this clinical efficacy lead Broxmeyer *et al.*[21] to suggest that a single collection of cord blood could be sufficient for adult transplantation.

EXPANSION OF CORD BLOOD STEM CELLS

In preclinical studies, cord blood transplantation (CBT) has rescued lethally irradiated mice and reconstituted their bone marrow.[22,23] After these leading experiments, clinical trials using cord blood transplantations have been applied to more than 2500 patients, mostly children, thus far. Insufficient number of stem cells obtained from a single cord blood has hampered extensive applications in adults, who required compilation of blood cells from several umbilical cords. Predictably, a primary goal in the field of cord blood transplantation has been *ex vivo* expansion and amplification of its stem cell content by various manipulations.

Similar to the case in bone marrow stem cells, expansion of cord blood stem cells by cocktails of growth factors has been attempted and reported with variable degrees of success.[24,25] These factors included cytokines like stem cell factor (SCF), FLT3L, thrombopoietin, and chemokines like IL-8, MIP1a, and VEGF, in addition to glycoaminoglycan.[26] Expanded cells following any of these protocols are assessed by both phenotypes and culture characteristics *in vitro,* by functional assays like colony-forming cell assay (CFC assay), and long-term culture–initiating cell assay (LTC-IC assay). Expansion of cord blood stem cells by co-culture on feeder layers of marrow stromal cells, or more recently, cord blood mesenchymal stem cells has also been reported.[27] Likewise, this co-stimulatory function of stromal cells has been shown *in vivo;* co-transplantation of mesenchymal stem cells has facilitated engraftment of cord blood stem cells in various mouse models.[28,29]

PLASTICITY OF CORD BLOOD CELLS

Several successful CBTs have been performed for both malignant and nonmalignant diseases of blood and other organs.[30] Diseases of the nervous system are an especially attractive target for stem cell therapy, since neurodegeneration is considered an end-stage illness and treatment for the most part is symptomatic. In addition, the discovery of neural stem cells and the potential of other stem cells to transdifferentiate into neural tissues have expanded neuroscience research in that direction. Studies that showed differentiation of embryonic stem (ES) cells *in vitro* into all neurons and glia[31] generated justifiable excitement about their therapeutic potential to replace degenerating neural cells. Ethical and medical concerns have directed stem cell research nationwide to alternative sources for plastic stem cells including HUCB stem cells. Many reports have followed from our laboratories and several others that show that HUCB stem cells could differentiate across tissue lineage boundaries into neural and other tissue lineages.

Sanchez-Ramos et al.[32] have demonstrated that the culture of mononuclear fraction of HUCB in a proliferating medium supplemented with all-*trans*-retinoic acid (RA) and nerve growth factor (NGF) promoted the expression Musashi-1 and TUJ-1 neural markers, and GFAP astrocyte marker (glial fibrillary acidic protein). In addition, mRNA for neuronal markers nestin and necdin was detected. Likewise, Ha et al.[33] have shown that HUCB cultured in beta-mercaptoethanol differentiated into neural phenotype as determined by positive immunocytochemical expression of neural nuclear antigen (NeuN), neurofilament, and GFAP, and by RT-PCR mRNA for nestin, neurofilament and microtubule-associated protein (MAP2). McGuckin et al.[34] have recently demonstrated that HUCB cells could expand in liquid culture supplemented with thrombopoietin, flt-3 ligand, and c-kit ligand (TPOFLK) into both hematopoietic and neuroglial progenitors.

EXPERIMENTS WITH SELECTED CORD BLOOD STEM CELLS

In most of the aforementioned studies, the mononuclear fraction of cord blood cells was used in the initial culture, without prior purification or selection of a precursor cell of interest. Whether neural differentiation of cord blood cells is the progeny of neural progenitors, or hematopoietic, or mesenchymal, or other stem cells within the cord blood graft was not clear. It became important to characterize a specialized cell fraction or population within the cord blood that is enriched for neural precursors for purposes of targeted therapy, genetic manipulations, and to further understand stem cell biology.

Bicknese et al.[35] purified a multipotent HUCB cell subset that is negative for the CD14 monocyte marker, and the CD34 hematopoietic progenitor marker. The culture was supplemented with basic fibroblast growth factor (bFGF) and human epidermal growth factor (hEGF). Immunohistochemistry and Western blot analysis showed differentiation into cells that expressed both GFAP and TUJ1 astrocyte and neural markers after 7 days in culture. In another study, a different HUCB cell fraction that is positive for both CD34 and the leukocyte marker CD45 was isolated by Buzanska et al.[36] by means of magnetic cell sorting. These clonic cells were, however, incapable of forming hematopoietic colonies. Upon culture in DMEM and hEGF, cells positive for nestin were produced. After further exposure to retinoic acid and BDNF, cells were immunoposive for TUJ1, MAP2, GFAP, and Gal-C (galacto-cerebroside) oligodendrocyte marker.

Studies of bone marrow stromal cells have shown that MSCs could be induced under conditions that increase intracellular levels of cAMP, to differentiate into cells of neural phenotype.[37] Under the appropriate culture conditions, both rodent and human marrow MSCs could be induced to differentiate into neural like cells.[38,39] Isolation of MSCs has not been as promising in cord blood as it is in bone marrow. Earlier experiments to separate MSCs from cord blood have either failed or obtained a low yield. However, further studies from various laboratories, including an ongoing study in our lab, have shown that MSCs could be isolated from cord blood, and could engraft immune-deficient mice.[40] In a recent study, Jeong *et al.*[41] isolated adherent cells expressing MSC-related antigens such as SH2, CD13, CD29, and AS-MA, from a mononuclear cell fraction of HUCB. After culture in neurogenic differentiation medium, both immunofluorescence and RT-PCR analyses indicated elevated expression of Tuj1, TrkA, GFAP and CNPases neural markers. Most of these studies agree that MSCs collected from cord blood are somewhat different in both structure, function, and abundance from bone marrow MSCs. Bieback *et al.*[42] have set strict criteria to obtain MSCs from cord blood, such as a time from collection to isolation of less than 15 hours, a net volume of more than 33 mL, and mononuclear cell count of at least 1×10^8 cells.

In view of available literature on cord blood stem cells, it is premature to define the cord blood cell fraction most enriched for neuroprogenitors. Further studies that include clonal analysis of transfected cells, neural and glial functional assays, and *in vivo* transdifferentiation are required before defining and isolating plastic cord blood cell populations. Most of the ongoing research is influenced by the bone marrow system, which may be similar to cord blood in terms of hematpoietic reconstitution, but significantly different in both cellular structure and content.

PRECLINICAL STUDIES UTILIZING HUCB CELLS

In vitro plasticity studies strongly suggest that cord blood stem cell therapy may represent a viable alternative for brain repair. Preliminary *in vivo* investigative studies in our laboratories examined homing, migration, and differentiation of HUCB cells into normal brains.[43] HUCB mononuclear fraction was transplanted into the subventricular zone of neonatal rat pups. Thirty days after transplantation, the pups were euthanized and brains dissected for human cells. HUCB cells were detected in the subventricular zone, the overlying cortex, and corpus callosum. Immunohistochemical phenotyping showed GFAP- and TUJ1-positive cells of donor HUCB origin in the developing brain, indicating differentiation into glial and neural phenotypes. The safety of this form of xenogenic transplantation was determined by absence of histological abnormalities or behavioral deficits in the transplanted rats. In both allo- and xenotransplants, however, the use of immune suppressive therapy enhanced, and at times was necessary, to maintain the survival of donor cells.[44]

In addition to the mononuclear fraction, purified populations of cord blood stem cells were also transplanted crossing the xenogenic barrier. A pluripotent, CD45⁻ population from HUCB, termed unrestricted somatic stem cells (USSCs), has been recently described by Kogler *et al.*[45] *In vitro*, USSCs differentiated into cells of multiple lineages including osteoblasts, chondroblasts, adipocytes, hematopoietic cells, and also astrocytes and neurons that express neurofilament, sodium channel protein, and various neurotransmitter phenotypes. When USSCs were transplanted into adult rat brain, human tau-positive cells with a typical neuron morphology were detected for up to 3 months post transplantation and showed neuron-like migratory activity.

Marrow stromal cell therapy showed promise in an ischemic cortex model where more marrow-derived cells were shown to migrate to the site of injury,[46] and to promote regeneration of the brain architecture in experimental autoimmune encephalitis.[47] In a promising clinical trial, *ex vivo* expansion of autologous mesenchymal stem cells and transplantation into the spinal cord of humans proved to be safe and well tolerated by ALS patients.[48]

HUCB CELL THERAPY FOR ISCHEMIC BRAIN DAMAGE

Stroke, one of the leading causes of death worldwide, is produced by focal ischemia to the brain and subsequent neural degeneration and damage. The basis for the use of cellular therapy in animal models for stroke is to stimulate neural regeneration and to limit further damage. The beneficial effect of HUCB transplantation in stroke-affected animals has challenged the dogma of neural rejuvenation and offered a viable system to analyze the cellular and molecular events involved in this regeneration.

There seems to be a consensus that ischemic injury in vital organs like the heart and the brain promotes cell migration to the site of injury to initiate the process of repair. Stimulation of endogenous stem cells that are activated and mobilized in response to various injuries seems to be an exciting strategy to promote endogenous repair of the adult CNS. However, the capacity of these progenitors to migrate and to differentiate into neural or glial cells differs according to the lesion type and the germinative zone from which they arise (reviewed by Picard-Riera *et al.*[49]). Studies in rodent models of stroke suggested that this process may be maintained by intrinsic stem cells that reside in the subventricular zone. Similar stem cells depicted as subventricular astrocytes have been recently proposed as a source for neural stem cells in adult humans.[50]

Ischemic neural injury stimulates inflammatory processes associated with recruitment and release of mediators crucial for initiating repair and supporting regeneration. This was demonstrated by *in vitro* studies using migration assays, in which extracts of ischemic tissues promoted migration of HUCB as compared to results in healthy controls.[51] Neurotransmitters involved in this inflammatory process included NGF (neurotrophic growth factor), BDNF (brain-derived neurotrophic factor), EGF (epidermal growth factor), FGF-2 (fibroblast growth factor-2), IGF, erythropoietin, and SCF (stem cell factor) (reviewed by Peterson[52]).

The evidence for the effect of HUCB transplantation on neural recovery after stroke first came from studies in Chopp's laboratories,[51,53] where rats were subjected to middle cerebral artery occlusion (MCAO) to induce focal ischemia-like pathology. Systemic delivery of HUCB via lateral tail-vein injection helped improvement in the transplanted animals and were detected in the affected cortex, subcortex, and striatum of the damaged brain. Immnuohistochemical phenotyping showed positive staining for neuronal markers (NeuN and MAP-2), astrocyte marker GFAP, and endothelial marker FVIII. Homing studies showed that tissue damage induced by traumatic brain injury stimulated migration of infused cord blood to the parenchyma of the affected brain tissue. Expression of neural and astrocyte markers was associated with functional improvements and reduction of motor and neuronal deficits.

Studies in stroke models from our laboratories have suggested that this improvement is linked to several factors including the site and extent of the neural damage, timing of the transplant, and the route of cellular administration. In the studies by Willing *et al.*,[54] stroke was similarly induced in rats by MCAO. Transplantation of HUCB into the striatum or the femoral veins of rats resulted in alleviated behavioral deficits. Nonetheless, this improvement was not associated with detection of donor HUCB cells in the brain by immunostaining methods. Cellular administration via the femoral vein was less invasive and associated with recovery of the forelimb. A subsequent study in our laboratories[55] has shown significant improvement in behavioral recovery 4 weeks after MCAO occlusion. Migration of HUCB was observed only in the injured hemisphere, and better recovery was correlated with higher doses of infused cord blood.

Another study by Saporta *et al.*[56] tested the effect of intravenous injection of cord blood on the recovery from spinal cord injury. HUCB was injected into rats 1 and 5 days after compression spinal cord injury. HUCB cells were localized around the site of injury, but not in the healthy spinal cord tissues. Behavioral open-field test scores were improved in the rat group undergoing transplantation 5 days after the injury. All of these studies have demonstrated that tissue injury is a critical factor in attracting donor cord blood cells and initiating the process of repair. Release of trophic factors at the site of injury may simultaneously promote selective migration of donor stem cells, and also accelerate healing and tissue repair.

Despite detection of donor cells at the site of injury, which cells within the cord blood graft contribute to the observed recovery and how this repair is achieved remain critical questions in neural transplantation. In a recent study, Borlongan *et al.*[57] were not able to detect HUCB in the brains of rats transplanted with low doses via the intravenous route after induction of stroke. Despite that a blood–brain barrier permeabilizer (mannitol) was co-infused with the cord blood cells, human cells were not engrafted. Reduced cerebral infarct size and increased levels of neuroprotectant factors led the investigators to suggest that the observed recovery after stroke was mediated by trophic factors and molecules induced by the cord blood cells, regardless of the availability of these cells at the site of the damage. A different mechanism was suggested by a study by Taguchi *et al.*,[58] who used the CD34+ cells within the cord blood to study the effect of this stem cell–enriched population on recovery from stroke. Immunocompromised mice underwent intravenous transplantation with CD34+ cells 48 hours after the induction of stroke. Recovery was associated with enhanced neovascularization on the borders of the ischemic zone. Interestingly, when this neovasculaization was suppressed by antiangiogenic agents, neurogenesis was impaired. This study suggests that systemic administration of cord blood CD34+ cells stimulated neovascularization, which in turn stimulated endogenous neurogenesis. Despite these encouraging data, the role of stem cell therapy in stroke is still elusive. Several factors like the type of injected cells (whole cord blood versus purified stem cells), the route of transplantation (systemic versus local, and intra-arterial versus intravenous), and most importantly, the window of time for successful therapy (less than 13 hours to up to 1 week) remains to be determined.

NEURODEGENERATIVE DISEASES

Neurodegenerative diseases include a group of brain disorders characterized by slow onset and progressive course of deteriorating neural functions. Nerve cell dysfunction caused by degeneration of specific brain regions affects memory and movement, usually in middle-aged and older populations. Several years usually elapse between the onset of the pathology and the clinical symptoms of disorders like Alzheimer's disease (AD), Parkinson's disease (PD), multiple sclerosis (MS), and amyotrophic lateral sclerosis (ALS). While advances in the treatment of these illnesses have dramatically improved the quality of life and elongated life span for afflicted patients, no cure is currently available. The impact of cellular therapy approaches on neurodegenerative diseases has been both encouraging and intangible. An important obstacle has been the scarcity of animal models to provide the appropriate *in vivo* opportunities to study and design more effective forms of therapy.

Amyotrophic lateral sclerosis (ALS) is a neurodegenerative disease caused by progressive degeneration of the motor neurons. Patients with ALS suffer weakness and progressive wasting and paralysis of the muscles that affect their ability to move, speak, swallow, and eventually breathe. Progressive paralysis could lead to death, usually within 5 years of disease onset. Currently, there is no cure for ALS, but treatment focuses on symptoms to improve the quality of life and delay complications. Recently, animal studies show that cell therapy could be a viable option to treatment of ALS by means of replacing diseased cells, stimulating neural cell regeneration, and delaying neural and motor atrophy.

The SOD-1 transgenic (B6SJL-TgN[SOD1-G93A]1GUR) mouse has a mutation of the human transgene, CuZn superoxide dismutase gene SOD1, which has been associated with amyotrophic lateral sclerosis. Ende et al.[59] and Chen et al.[60] attempted the cell therapy approach by transplanting large doses of HUCB mononuclear cells into SOD mice by intravenous administration. HUCB transplantation caused considerable delay in the onset of symptoms and death of the ALS mouse model.

In a recent study in our laboratories, Garbuzova-Davis et al.[61] have administered MNC fraction of HUCB via jugular vein injection into a presymptomatic G93A ALS mouse model prior to the onset of behavioral impairments. Significant benefits were observed in hind-limb extension and gait. Mice undergoing transplantation maintained their weight better and had a significantly longer life span than diseased nontransplanted mice. Cord blood graft has provided protection of motor neurons and perhaps replacement of damaged neurons in ALS-affected mice. The mechanism of repair after cord blood transplantation is, however, not well understood. FIGURE 1 shows immunophenotyping of HUCB in the brain after intravenous injections. Donor cord blood cells, detected in the parenchyma of the brain and spinal cord (FIG. 2), were positive for neural and astrocyte phenotype markers (TUJ1 and GFAP). Donor cells were also detected in the spleen, kidneys, liver, lungs, and heart.

CLINICAL STUDIES WITH CORD BLOOD

An expanding list of disorders currently treated with cord blood transplantation include hemoglobinopathies like sickle cell anemia and thalassemia, leukodystro-

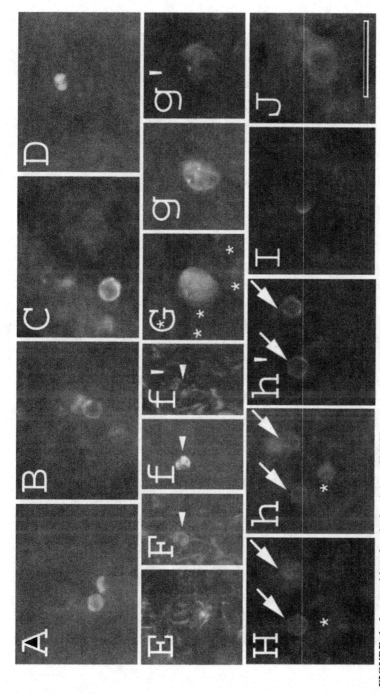

FIGURE 1. Immunohistochemical staining of hUCB cells in the brain. Most cells were found in blood vessels (**A,B,C**) in many structures of the brain. They were (**A**) HuNu/CD45-positive (orange/red, double staining) or (**B,C**) only HuNu-positive (green). Some cells were found (**C**) parenchymally at a distance from the blood vessel in (**D**) corpus callosum, (**E**) ventral tegmental area, (**F**) pons, (**G**) frontal cortex, (**H**) olfactory bulb,(**I**) lateral olfactory tract, and (**J**) hippocampus. The cells were also found in the striatum, cerebellar lobules, ventromedial thalamus, and spinal trigeminal area. Some cells were double-stained with HuNu and GFAP (**F**) and TuJ1 (**H,I,J**). The nuclei in **H** and **J** are shown with DAPI. Scale bar in **A, C, D, E, F, G, H, I,** and **J** is 25 mm. Scale bar in **B** is 50 mm. [*Illustration is in color online.*]

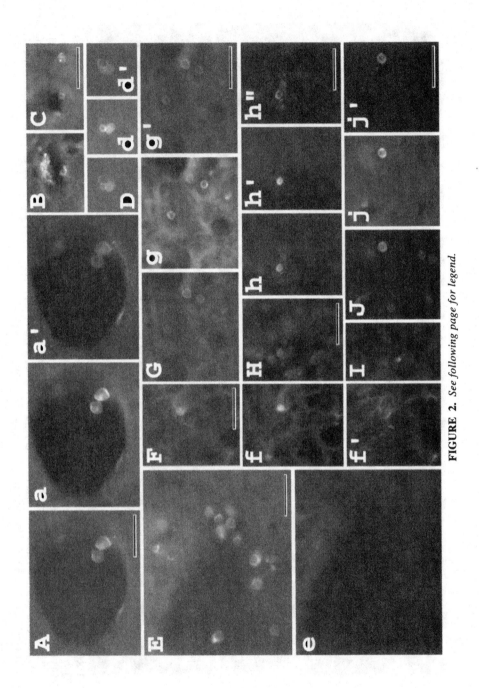

FIGURE 2. *See following page for legend.*

phies, severe combined immune deficiency, aplastic anemia, Fanconi's anemia, glycogen storage diseases like Hurler's syndrome and the Hunter syndrome, erythrocyte enzyme deficiencies and errors of metabolism. Additionally, cord blood cell therapy was promising in the treatment of acute lymphoblastic leukemia[62,63] and acute myeloid leukemia.[64] One hundred per cent donor engraftment and 5 years' disease-free survival was accomplished in a 2-month old infant suffering from beta-thalassemia after transplantation of partially MHC-matched HUCB from an unrelated donor.[65]

Progress of cell therapy applications for neurological disorders has been slower, Wenger *et al.*[66] have utilized cord blood transplantation to treat Krabbe's disease. This enzymatic disorder is caused by deficiency of galactocerebrosidase (CAL-C), which results in deficiency of myelin formation in both the central and the peripheral nervous system. Prognosis of this disease is grave in infants, but better in older patients. HUCB transplantation lessened the disease manifestations, but a cure was not achieved.

Another metabolic disease that seriously affects the nervous system is Hurler's syndrome. It is a severe form of mucopolysaccharidosis type I that affects children, and causes progressive deterioration of the central nervous system, which leads to death. In a study by Staba *et al.*[67] cord blood transplantation from unrelated donors was used to treat young children with Hurler's syndrome. A myeloablative preparative regimen that did not involve total-body irradiation was followed by cord blood transplantation. The children showed a high survival rate and were durably engrafted with donor cord blood.

CONCLUSION AND FUTURE PERSPECTIVES

Stem cell therapy for otherwise fatal diseases has greatly advanced since the first bone marrow transplant for severe combined immune deficiency.[68] Manipulation of the marrow stem cell graft by deleting mature GVHD-causing cells and enriching for the hematopoietic stem cell populations, and improvement of the myeloablative conditioning therapy prior to the transplant have considerably enhanced the outcome of stem cell therapy.[69] While autologous stem cell transplants provide the best engraftment outcome, allogeneic transplantation proved superior in delaying relapse in malignancies and in experimental autoimmunities.

FIGURE 2. Immunohistochemical staining of hUCB cells in the lumbar spinal cord. (**A,B**) Double-labeled HuNu/CD45 hUCB cells are found in or (**C**) outside the blood vessels (HuNu, green; CD45, red/orange). (**D**) Some cells in parenchymal location were negative for CD45 and only stained with HuNu (green). (**E**) Double-labeled cells expressing GFAP (red) and HuNu (green) or (**F**) surrounded by astrocytes (*arrowhead*). (**f**) same view as **F** with fluorescein and rhodamine (**f'**) filters. (**G**) Nestin expression (red) in double-labeled cell with HuNu (green). Stars indicate nuclei (DAPI) of mouse cells in tissue. (**g**) same view as **G** with fluorescein and (**g'**) rhodamine filters. (**H,I,J**) Cells double-labeled for TuJ1 (red/orange, *arrows*) and HuNu (green). (**h**) same view as **H** with fluorescein and (**h'**) rhodamine filters. The nuclei in **H** are shown with DAPI and stars (**H**) and (**h**) indicates negative stain for TuJ1. NOTE: Nestin-positive and TuJ1-positive cells differ morphologically from any administered hUCB cells identified within the spinal cord. Scale bar in **A–J** is 25 mm. [*Illustration is in color online.*]

Human umbilical cord blood is a highly promising source for cell therapy in a variety of diseases currently treated with bone marrow transplantation. This promise is particularly valuable for patients suffering from neural disorders for which no cure is available. *In vitro* studies that showed plasticity of HUCB cells, and *in vivo* studies that achieved not only delay or halt of neural degeneration, but also active restoration of neural functions, have created justifiable excitement in neural research. Unique qualities of HUCB like availability, immature cellular phenotype, enrichment for hematopoietic progenitors, plasticity, and lower incidence of graft versus host disease all appropriate its use as an ideal source for cell therapy (reviewed by Newman *et al.*[70]).

HUCB transplantation is a particularly attractive strategy for neurological disorders, because of the grave prognosis and current absence of a cure for most of these diseases, and the promising data from basic research and preclinical studies. Many practical factors influence the outcome of the therapy with CBT. For example the route of cord blood infusion is particularly critical in the CNS when we consider the blood–brain barrier. In animal models, intra-bone transplantation of cord blood has shown higher seeding efficiency than has the intravenous route.[71] This strategy could be especially beneficial when the number of cord blood stem cells is limited, or when direct delivery of stem cells to the site of damage is believed to initiate earlier repair within the critical hours after the ischemic or traumatic damage.

The abundance of *in vitro* data showing differentiation of various cord blood stem cells into neural and glial cells has tempted researchers to suggest that cord blood–induced brain repair is mediated by a transdifferentiation process. Plasticity of stem cells, however, has been a subject of intense debate.[72] Despite evidence for neural differentiation both *in vitro* and *in vivo*, limited evidence suggests that this phenotypic delineation involves functional neural cells. This lack of functional assays has directed most preclinical studies to gauge improvement by behavioral testing.

The promising data of improved behavior, delayed disease onset, and prolonged survival, coupled with an immense desire to find a cure for debilitating and devastating CNS diseases stimulated a renewed interest in stem cell therapy for neural disorders. The critical question to be pursued by researchers is what possible mechanisms are involved in neural repair by cell therapy. Replacement of diseased cells with functional "new" stem cells may be an accepted resolution for cases like leukemia cured by bone marrow transplantation; however, limited evidence in neuroscience studies suggests that this is the case.

While pursuing the inevitable leap of stem cell therapy from the bench top to the clinic, many questions need continued investigation, for example:

- What fraction of cord blood cells provides the maximum benefit with the fewest side effects? Are purified populations of stem cells superior to unseparated or mononuclear cell fractions?

- What is the most efficient way to deliver HUCB stem cells? Is local implantation superior to systemic injection?

- What is the optimal dose of HUCB cells, and are multiple injections required to maintain the desired therapeutic effect?

- How could the stem cell graft be manipulated so that minimal immune suppression is required without graft rejection?

- Which cells within the cord blood graft induce GVHD, and how could the graft be manipulated to deplete such undesired populations?

- How do growth factors and cellular mediators produced by the cord blood cells affect the progress of disease and how do these factors influence the production of intrinsic neurotransmitters and mediators?

- Which diseases of the CNS are most promising as targets for stem cell therapy and what pathology associated with these disorders permits a curative effect mediated by stem cells?

- How does local pathology influence migration of donor stem cells to the site of injury, and promote differentiation into specialized, functional neural cells?

- How are behavioral functions affected by stem cell therapy and how do these functions relate to neural recovery?

And last but not least,

- How do cord blood cells initiate and maintain neural repair?

Modulation of the host immune responses, stimulation of endogenous host stem cells, and production of various neural mediators and epigenic growth factors have all been suggested as mechanisms to be explored by scientists.

For patients and clinicians, however, the hope sparked by case reports of neural recovery after stem cell transplants makes the cell therapy approach for neurological disorders well worthy of attention.

ACKNOWLEDGMENTS

We appreciate the excellent assistance of Donna C. Morrison.

DISCLOSURE

Paul R. Sanberg is chairman and co-founder, and Alison Willing, Svitlana-Garbuzova-Davis, and Samuel Saporta are consultants of Saneron CCEL Therapeutics, Inc.

REFERENCES

1. BADEN, L. & RUBIN, R.H. 2004. Infection in hematopoietic stem cell transplant recipients. *In* Stem Cell Transplantation for Hematologic Malignancies. R.A. SoifferEd. : 237–258. Humana Press. Totowa, NJ.
2. BROXMEYER, H.E., G.W. DOUGLAS, G. HANGOC, G., *et al.* 1989. Human umbilical cord blood as a potential source of transplantable hematopoietic stem/progenitor cells. Proc. Natl. Acad. Sci. USA **86:** 3828–3832.
3. MAHMOUD, H., O. FAHMY, A. KAMEL, *et al.* 1999. Peripheral blood versus bone marrow as a source for allogeneic hematopoietic stem cell transplantation. Bone Marrow Transplant. **5:** 28–35.
4. FERRARA, J.L., R. LEVY & N.J. CHAO. 1999. Pathophysiologic mechanisms of acute graft versus host disease. Biol. Blood Marrow Transplant. **5:** 347–356.

5. PARKMAN, R. Chronic graft-versus-host disease. 1998. Curr. Opin. Hematol. **5:** 22–25.
6. GLUCKMAN, E. & V. ROCHA. 2004. Cord blood transplant: strategy of alternative donor search. Springer Semin. Immunopathol. **26:** 143–154.
7. WAGNER, J.E., N.A. KERNAN, M. STEINBUCH, *et al.* 1995. Allogeneic sibling umbilical-cord-blood transplantation in children with malignant and non-malignant disease. Lancet **346:** 214–219.
8. ROCHA, V., J. CORNISH, E.L. SIEVERS, *et al.* 2001. Comparison of outcomes of unrelated bone marrow and umbilical cord blood transplants in children with acute leukemia. Blood **97:** 2962–2971.
9. VAZIRI, H., W. DRAGOWSKA, R.C. ALLSOPP, *et al.* 1994. Evidence for a mitotic clock in human hematopoietic stem cells: loss of telomeric DNA with age. Proc. Natl. Acad. Sci. USA **91:** 9857–9860.
10. RIDSON, G., J. GADDY, & H.E. BROXMEYER. 1994. Allogeneic responses of human umbilical cord blood. Blood Cells **20:** 560–565.
11. RONCAROLO, M., M. BIGLER, E. CIUTI, *et al.* 1994. Immune responses by cord blood cells. Blood Cells **20:** 573–586.
12. BRACHO, F., C. VAN DE VEN, E. AREMAN, *et al.* 2003. A comparison of ex vivo expanded DCs derived from cord blood and mobilized adult peripheral blood plastic-adherent mononuclear cells: decreased alloreactivity of cord blood DCs. Cytotherapy **5:** 349–361.
13. WILSON, C.B., J. WESTALL, L. JOHNSTON, *et al.*1986. Decreased production of interferon-gamma by human neonatal cells. Intrinsic and regulatory deficiencies. J. Clin. Invest. **77:** 860–867.
14. KNUTSEN, A.P. & D.A. WALL. 1999. Kinetics of T-cell development of umbilical cord blood transplantation in severe T-cell immunodeficiency disorders. J. Allergy Clin. Immunol. **103:** 823–832.
15. TRAGGIAI, E., L. CHICHA, L. MAZZUCCHELLI, *et al.* 2004. Development of a human adaptive immune system in cord blood cell-transplanted mice. Science **304:** 104–107.
16. NIEHUES, T., V. ROCHA, A.H. FILIPOVICH, *et al.* 2001. Factors affecting lymphocyte subset reconstitution after either related or unrelated cord blood transplantation in children: a Eurocord analysis. Br. J. Haematol. **114:** 42–48.
17. GOGGINS, T.F. & N.J. CHAO. 2004. Umbilical cord hematopoietic stem cell transplantation. *In* Stem Cell Transplantation for Hematologic Malignancies. R.J. Soiffer, Ed. : 391–416.Humana Press. Totawa, NJ.
18. THEILGAARD-MONCH, K., K. RAASCHOU-JENSEN, H. PALM, *et al.* 2001. Flow cytometric assessment of lymphocyte subsets, lymphoid progenitors, and hematopoietic stem cells in allogeneic stem cell grafts. Bone Marrow Transplant **28:** 1073–1082.
19. HAO, Q.L., A.J. SHAH, F.T. THIEMANN, *et al.* 1995. A functional comparison of CD34$^+$ CD38$^-$ cells in cord blood and bone marrow. Blood **86:** 745–3753.
20. GLUCKMAN, E., V. ROCHA & C. CHASTANG. 1999. Peripheral stem cells in bone marrow transplantation. Cord blood stem cell transplantation. Baillieres Best Pract. Res. Clin. Haematol. **12:** 279–292.
21. BROXMEYER, H.E., G. HANGOC, S. COOPER, *et al.* 1992. Growth characteristics and expansion of human umbilical cord blood and estimation of its potential for transplantation in adults. Proc. Natl. Acad. Sci. USA **89:** 4109–4113.
22. BROXMEYER, H.E., J. KURTZBERG, E. GLUCKMAN, *et al.* 1991. Umbilical cord blood hematopoietic stem and repopulating cells in human clinical transplantation. Blood Cells **17:** 313–329.
23. BROXMEYER, H.E. 1996. Primitive hematopoietic stem and progenitor cells in human umbilical cord blood: an alternative source for transplantable cells. Cancer Treat. Res. **84:** 139–148.
24. YAO, C.L., L.M CHU, T.B. HSIEH & S.M. HWANG. 2004. A systematic strategy to optimize ex vivo expansion medium for human hematopoietic stem cells derived from umbilical cord blood mononuclear cells. Exp. Hematol. **32:** 720–727.
25. GLUCKMAN, E. 2004. Ex vivo expansion of cord blood cells. Exp. Hematol. **32:** 410–412.
26. WAGNER, J.E. & C.M VERFAILLIE. 2004. Ex vivo expansion of umbilical cord blood hemopoietic stem and progenitor cells. Exp. Hematol. **32:** 412–413.

27. ZHANG, Y., C. LI, X. JIANG, *et al.* 2004. Human placenta-derived mesenchymal progenitor cells support culture expansion of long-term culture-initiating cells from cord blood CD34+ cells. Exp. Hematol. **32:** 657–664.
28. ANGELOPOULOU, M., E. NOVELLI & J.E. GROVE. 2003. Cotransplantation of human mesenchymal stem cells enhances human myelopoiesis and megakaryocytopoiesis in NOD/SCID mice. Exp. Hematol. **31:** 413–420.
29. NOORT, W.A., A.B. KRUISSELBRINK, P.S. IN'T ANKER, *et al.* 2002. Mesenchymal stem cells promote engraftment of human umbilical cord blood derived CD34+ cells. Exp. Hematol. **8:** 870–878.
30. LU, L., R.N. SHEN & H.E. BROXMEYER. 1996. Stem cells from bone marrow, umbilical cord blood and peripheral blood for clinical application: current status and future application. Crit. Rev. Oncol. Hematol. **22:** 61–78.
31. MANSERGH, F.C., M.A. WRIDE & D.E. RANCOURT. 2000. Neurons from stem cells: implications for understanding nervous system development and repair. Biochem. Cell. Biol. **78:** 613–628.
32. SANCHEZ-RAMOS, J.R., S. SONG, S.G. KAMATH, *et al.* 2001. Expression of neural markers in human umbilical cord blood. Exp. Neurol. **171:** 109–115.
33. HA, Y., J.U. CHOI, D.H. YOON, *et al.* 2001. Neural phenotype expression of cultured human cord blood cells in vitro. Neuroreport **12:** 3523–3527.
34. MCGUCKIN, C.P., N. FORRAZ, Q. ALLOUARD & R. PETTENGELL. 2004. Umbilical cord blood stem cells can expand hematopoietic and neuroglial progenitors in vitro. Exp. Cell Res. **295:** 350–359.
35. BICKNESE, A.R., H.S. GOODWIN, C.O. QUINN, *et al.* 2002. Human umbilical cord blood cells can be induced to express markers for neurons and glia. Cell Transplant. **11:** 261–264.
36. BUZANSKA, L., E.K. MACHAJ, B. ZABLOCKA, *et al.* 2002. Human cord blood-derived cells attain neuronal and glial features in vitro. J. Cell Sci. **115:** 2131–2138.
37. DENG, W., M. OBROCKA, I. FISCHE & D.J. PROCKOP. 2001. *In vitro* differentiation of human marrow stromal cells into early progenitors of neural cells by conditions that increase intracellular cyclic AMP. Biochem. Biophys. Res. Commun. **282:** 48–152.
38. SANCHEZ-RAMOS, J., S. SONG, F. CARDOZO-PELAEZ, *et al.* 2000. Adult bone marrow stromal cells differentiate into neural cells in vitro. Exp. Neurol. **164:** 247–256.
39. WOODBURY, D., E.J. SCHWARZ, D.J. PROCKOP & L.B. BLACK. 2000. Adult rat and human bone marrow stromal cells differentiate into neurons. J. Neurosci. Res. **61:** 364–670.
40. ERICES, A.A., C.I. ALLERS, P.A. CONGET, *et al.* 2000. Human cord blood-derived mesenchymal stem cells home and survive in the marrow of immunodeficient mice after systemic infusion. Cell Transplant. **12:** 555–561.
41. JEONG, J.A., E.J. GANG, S.H. HONG, *et al.* 2004. Rapid neural differentiation of human cord blood-derived mesenchymal stem cells. Neuroreport **15:** 1731–1734.
42. BIEBACK, K., S. KERN, H. KLUTER & H. EICHLER. 2004. Critical parameters for the isolation of mesenchymal stem cells from umbilical cord blood. **22:** 625–634.
43. ZIGOVA, T., S. SONG, A.E. WILLING, *et al.* 2002. Human umbilical cord blood cells express neural antigens after transplantation into the developing rat brain. Cell Transplant. **11:** 265–74.
44. IRONS, H., J.G. LIND, C.G. WAKADE, *et al.* 2004. Intracerebral xenotransplantation of GFP mouse bone marrow stromal cells in intact and stroke rat brain: graft survival and immunologic response. Cell Transplant. **13:** 283–294.
45. KOGLER, G., S. SENSKEN, J.A. AIREY, *et al.* 2004. A new human somatic stem cell from placental cord blood with intrinsic pluripotent differentiation potential. J. Exp. Med. **200:** 123–135.
46. EGLITIS, M.A., D. DAWSON, K.W. PARK & M.M. MOURADIAN. 1999. Targeting of marrow-derived astrocytes to the ischemic brain. Neuroreport **26:** 1289–1292.
47. FLUGEL, A., M. BRADL, G.W. KREUTZBERG & M.B. GRAEBER. 2001. Transformation of donor-derived bone marrow precursors into host microglia during autoimmune CNS inflammation and during the retrograde response to axotomy. J. Neurosci. Res. **66:** 74–82.

48. MAZZINI, L., F. FAGIOLI, R. BOCCALETTI, et al. 2003. Stem cell therapy in amyotrophic lateral sclerosis: a methodological approach in humans. Amyotroph. Lateral. Scler. Other Motor Neuron Disord. **4:** 158–161.
49. PICARD-RIERA, N., B. NAIT-OUMESMAR, B. & A. EVERCOOREN. 2004. Endogenous adult neural stem cells: limits and potential to repair the injured central nervous system. J. Neurosci. Res. **76:** 223–231.
50. SANAI, N., A.D. TRAMONTIN, A. QUINONES-HINOJOSA, et al. 2004. Unique astrocyte ribbon in adult human brain contains neural stem cells but lacks chain migration. Nature. **427:** 685–686.
51. CHEN, J., P.R. SANBERG, Y. LI, et al. 2001. Intravenous administration of human umbilical cord blood reduces behavioral deficits after stroke in rats. Stroke **32:** 2682–2688.
52. PETERSON, D.A. 2004. Umbilical cord blood cells and brain stroke injury: bringing in fresh blood to address an old problem. J. Clin. Invest. **114:** 312–314.
53. LU, D., P.R. SANBERG, A. MAHMOOD, et al. 2002. Intravenous administration of human umbilical cord blood reduces neurological deficit in the rat after traumatic brain injury. Cell Transplant. **11:** 275–281.
54. WILLING, A.E., J. LIXIAN, M. MILLIKEN, et al. 2003. Intravenous versus intrastriatal cord blood administration in a rodent model of stroke. J. Neurosci. Res.**73:** 296–307.
55. VENDRAME, M., J. CASSADY, J. NEWCOMB, et al. 2004. Infusion of human umbilical cord blood cells in a rat model of stroke dose-dependently rescues behavioral deficits and reduces infarct volume. Stroke **35:** 2390–2395.
56. SAPORTA, S., J.J. KIM, A.E. WILLING, et al. 2003. Human umbilical cord blood stem cells infusion in spinal cord injury: engraftment and beneficial influence on behavior. J. Hematother. Stem Cell Res. **12:** 271–278.
57. BORLONGAN, C.V., M. HADMAN, C. DAVIS-SANBERG, & P.R. SANBERG. 2004. Central nervous system entry of peripherally injected umbilical cord blood cells is not required for neuroprotection in stroke. Stroke **35:** 2385–2389.
58. TAGUCHI, A., SOMA, T., TANAKA, H., et al. 2004. Administration of CD34[+] cells after stroke enhances neurogenesis via angiogenesis in a mouse model. J. Clin. Invest. **114:** 330-338.
59. ENDE, N., F. WEINSTEIN, R. CHEN & M. ENDE. 2000. Human umbilical cord blood effect on SOD mice (amyotrophic lateral sclerosis). Life Sci. **67:** 53–59.
60. CHEN, R. & N. ENDE. 2000. The potential for the use of mononuclear cells from human umbilical cord blood in the treatment of amyotrophic lateral sclerosis in SOD1 mice. J. Med. **31:** 21–30.
61. GARBUZOVA-DAVIS, S., A.E. WILLING, T. ZIGOVA, et al. 2003. Intravenous administration of human umbilical cord blood cells in a mouse model of amyotrophic lateral sclerosis: distribution, migration, and differentiation. J. Hematother. Stem Cell Res. **12:** 255–570.
62. LUAN, Z., S.X. XU & N.H. WU. 2004. [Treatment of refractory and relapsed childhood acute leukemia by HLA-mismatched unrelated umbilical cord blood transplantation.] Zhonghua Er Ke Za Zhi. **42:** 535.
63. WANG, L., X.J. HUANG, & X.X. CHEN. 2004. [Successful treatment of one case acute lymphoblastic leukemia by HLA-mismatched unrelated umbilical cord blood transplantation.] Zhonghua Er Ke Za Zhi **42:** 552.
64. BERGER, M., E. VASSALLO, F. NESI, et al. 2004. Successful unrelated cord blood transplantation following reduced-intensity conditioning for refractory acute myeloid leukemia. J. Pediatr. Hematol. Oncol. **26:** 98–100.
65. HALL, J.G., P.L. MARTIN, S. WOOD & J. KURTZBERG. 2004 Unrelated umbilical cord blood transplantation for an infant with beta-thalassemia major. J. Pediatr. Hematol. Oncol. **26:** 382–385.
66. WENGER, D.A., M.A. RAFI, P. LUZI, P., et al. 2000. Krabbe disease: genetic aspects and progress toward therapy. Mol. Genet. Metab. **70:** 1–9.
67. STABA, S.L., M.L. ESCOLAR, M. POE, et al. 2004. Cord-blood transplants from unrelated donors in patients with Hurler's syndrome. N. Engl. J. Med. **350:** 1960–1969.
68. GOOD, R.A. 2002. Cellular immunology in a historical perspective. Immunol. Rev. **185:** 136–158.

69. EL-BADRI, N.S., A. MAHESHWAI & P.R. SANBERG. 2004. Mesenchymal stem cells in autoimmune disease. Stem Cell Dev. **13:** 463–472.
70. NEWMAN, M.B., C.D. DAVIS, C.V. BORLONGAN, *et al.* 2004. Transplantation of human umbilical cord blood cells in the repair of CNS diseases. Expert Opin. Biol. Ther. **4:** 121–30.
71. CASTELLO, S., M. PODESTA, V.G. MENDITTO, *et al.* 2004. Intra-bone marrow injection of bone marrow and cord blood cells: an alternative way of transplantation associated with a higher seeding efficiency. Exp. Hematol. **32:** 782–787.
72. VERFAILLIE, C.M., M.F. PERA & P.M. LANSDORP. 2002. Stem cells: hype and reality. Hematology (Am. Soc. Hematol. Educ. Program) :369–391.

Infusion of Human Umbilical Cord Blood Ameliorates Neurologic Deficits in Rats with Hemorrhagic Brain Injury

ZHENHONG NAN,[a] ANDREW GRANDE,[a] CYNDY D. SANBERG,[d]
PAUL R. SANBERG,[e] AND WALTER C. LOW[a,b,c]

[a]Department of Neurosurgery, and [b]Graduate Program in Neuroscience, and [c]Stem Cell Institute, University of Minnesota, Medical School, Minneapolis, Minnesota 55455, USA

[d]Saneron CCEL Therapeutics, Inc., Tampa, Florida 33612, USA

[e]Department of Neurosurgery, and Center of Excellence for Aging and Brain Repair, University of South Florida, Tampa, Florida, USA

ABSTRACT: Umbilical cord blood is a rich source of hematopoietic stem cells. It is routinely used for transplantation to repopulate cells of the immune system. Recent studies, however, have demonstrated that intravenous infusions of umbilical cord blood can ameliorate neurologic deficits associated with ischemic brain injury in rodents. Moreover, the infused cells penetrate into the parenchyma of the brain and adopt phenotypic characteristics typical of neural cells. In the present study we tested the hypothesis that the administration of umbilical cord blood can also diminish neurologic deficits caused by intracerebral hemorrhage (ICH). Intracerebral hemorrhage is a major cause of morbidity and mortality, and at the present time there are no adequate therapies that can minimize the consequences of this cerebrovascular event. ICH was induced in rats by intrastriatal injections of collagenase to cause bleeding in the striatum. Twenty-four hours after the induction of ICH rats received intravenous saphenous vein infusions of human umbilical cord blood (2.4×10^6 to 3.2 to 10^6 cells). Animals were evaluated using a battery of tests at day 1 after ICH, but before the administration of umbilical cord blood, and at days 7, and 14 after ICH (days 6 and 13, respectively, after cord blood administration). These tests included a neurological severity test, a stepping test, and an elevated body-swing test. Animals with umbilical cord blood infusions exhibited significant improvements in (1) the neurologic severity test at 6 and 13 days after cord blood infusion in comparison to saline-treated animals ($P < 0.05$); (2) the stepping test at day 6 ($P < 0.05$); and (3) the elevated body-swing test at day 13 ($P < 0.05$). These results demonstrate that the administration of human umbilical cord blood cells can ameliorate neurologic deficits associated with intracerebral hemorrhage.

KEYWORDS: umbilical cord blood; intracerebral hemorrhage; functional recovery

Address for correspondence: Walter C. Low, Ph.D., Department of Neurosurgery, University of Minnesota Medical School, 2001 Sixth St., S.E., Minneapolis, MN 55455. Voice: 612-626-9200; fax: 612-626-9201.
lowwalt@umn.edu

Ann. N.Y. Acad. Sci. 1049: 84–96 (2005). © 2005 New York Academy of Sciences.
doi: 10.1196/annals.1334.009

INTRODUCTION

Intracerebral hemorrhage (ICH) is a major cause of mortality and morbidity. Nearly 20% of patients who have an ICH die within 3 days after diagnosis.[1] The two-year survival rate is 65%, and the outlook for stroke survivors is poor, with the majority having to endure life-long debilitating neurologic impairments. The putaminal component of the basal ganglia is the most common site for intracerebral hemorrhage and accounts for about one-third to one-half of all ICHs.[2–4] Patients with putaminal hemorrhage present with contralateral hemiplegia, sensory impairment, and either aphasia with dominant hemisphere involvement or contralateral neglect with involvement of the nondominant hemisphere.[5–7] At the present time there is no therapy to restore function in patients with ICH.

Umbilical cord blood (UCB) is conventionally used as an alternative source of cells for bone marrow transplantation in a variety of hematologic disorders and following high-dose chemotherapy for the treatment of cancer. Recently, UCB has been shown to be effective in ameliorating neurologic deficits associated with ischemic and traumatic brain injuries in rodents. In rats with ischemic brain injury induced by unilaterally occluding the middle cerebral artery, intravenous injections of human UCB 24 hours after ischemic injury resulted in the amelioration of neurologic deficits.[8] Histologic examination of brain tissue in animals with UCB injections revealed cells that were recognized by antibodies against human nuclei (MAB1281). MAB1281-positive cells also expressed GFAP and NeuN, markers for astrocytes and neurons, respectively. In rats with traumatic head injury, intravenous injections of UCB also resulted in the amelioration of behavioral deficits as revealed by rotarod tests and a neurologic severity score test.[9] The reduction in behavioral deficits in these animals was also associated with the presence of human cells that expressed markers for astrocytes and neurons. These studies suggest that certain cells contained within umbilical blood have the ability to differentiate into neural cells that might account for the restorative effects observed in animals with UCB injections. In addition, other cytokines and growth factors may also participate in functional recovery.

Umbilical cord blood is a rich source of stem/progenitor cells. Cord blood has been shown to contain hematopoietic stem cells,[10] mesenchymal stem cells,[11] and endothelial precursor cells.[12] At the present time it is not known whether any of these stem/progenitor cells contribute to the amelioration of neurologic deficits seen in animals treated with UCB. It is possible that a combination of cells within UCB contributes to functional restitution after transplantation in ischemic and traumatic brain injury. The therapeutic effects of human UCB has yet to be evaluated in hemorrhagic brain injury.

An experimental model of ICH using intraparenchymal collagenase has been reported by Rosenberg *et al.*[13] Collagenases are proteolytic enzymes that are present within cells in an inactive form and are secreted at sites of inflammation by mononuclear cells.[14,15] Cerebral tissue contains collagen in the basal lamina of blood vessels, and the direct injection of collagenase causes the local destruction of blood vessels. The sequence of pathologic changes such as leukocytic infiltration, cystic formation, and the appearance of erythrocytes within the parenchyma are reproducible and mimic the ICH features seen in humans. Assessment of motor function in these animals revealed deficits in grasping, beam walking, and spontaneous circling.

These results suggest that the rodent model of collagenase-induced ICH replicates many of the neuropathologic and neurochemical sequelae seen in spontaneous ICH. The present study examines the effects of human UCB infusions in this rodent model of ICH.

MATERIAL AND METHODS

Animal Model of Intracerebral Hemorrhage

The animal model for intracerebral hemorrhage is that described by Chesney et al.[16] Sprague Dawley rats, ranging in body weight from 200–225 g, were used for making intracranial hemorrhage. Under ketamine/xylazine anesthesis (1 mL/kg i.p.), each rat was set in a Kopf stereotaxic headholder, and the scalp was incised in the midline to expose the skull. A hole was made 0.4 mm anterior and 3.2 mm lateral to bregma. With a 10-µL Hamilton syringe, 0.5 unit of type VII collagenase (Sigma Chemical Co.) in 1 µL saline solution was injected into right striatum (coordinates: AP 0.4, ML 3.2, DV 5.0 mm with the tooth bar set at 0 mm) over a 2-minute period. The needle was then slowly withdrawn and the wound closed. Cyclosporin A was given daily (10 mg/kg, i.m.) to prevent the rejection of the human cells by the rats' "immune system."

Preparation of Human Umbilical Cord Blood Cells

On the day of cord blood infusion, vials containing human UCB (Saneron CCEL Therapeutics, Inc.) were taken from liquid nitrogen, rinsed, and centrifuged, and each vial was made into a final suspension (0.5 mL in saline) containing 2.4×10^6–3.2×10^6 mononucleated cells. Cell viability was 75%–90% as determined by trypan blue exclusion.

Administration of Umbilical Cord Blood Cells

On the first day after intracerebral hemorrhage, all rats were evaluated using a battery of behavioral tests that included a limb placement test, a stepping test, and an elevated body-swing test (EBST). Animals were then randomized into two groups: group 1 received UCB by intravenous (i.v.) administration; group 2 received i.v. saline solution. On the 7th day, all rats were tested as described below. On the 14th day, these tests were repeated, and all rats were anesthetized and perfused transcardially (PBS and 4% formaldehyde in sequence). Brains were collected, immersed in 4% para-formaldehyde for 24 hours, and then in 15% sucrose for at least 2 days until the brains sank. Brains were then cut surgically 30 µm in thickness using a frozen sliding microtome.

Limb-Placement Test

The limb-placement protocol is modified from the report by De Ryck.[17] The scoring system consists of 6 tests to evaluate forelimb and hindlimb function on each side of the body. The highest score of 16 is typically given to normal rats. For each test a score of 0 is given for a limb in which there was no placing response; 1 for incomplete or delayed (>2 seconds) limb placing; 2 for immediate and complete limb placing. The description of each test in more detail follows.

Test 1: The rat is held in mid-air by its tail. In this position its forelimbs are in the tucked position. The rat is then slowly lowered toward the surface of a counter top. As the animal's body approaches the counter top it is suspended 10 cm above the surface. In this position normal rats extend both forelimbs. In contrast, animals with unilateral hemorrhagic injury are unable to extend the affected contralateral forelimb.

Test 2: The rat is held around its body with its forepaws placed on the surface of a counter top with the animal's head raised 45° upward. Normal rats can extend their forepaws and maintain contact with the counter top. In contrast, rats with a unilateral stroke will lose contact of the contralateral forepaw.

Test 3: The rat is held around the body with forelimbs freely suspended. The animals are oriented so that they are facing a counter top, and then slowly moved towards its edge in a horizontal direction. Normal rats will extend both forepaws directly, and place them on the top edge of the counter top.

Test 4: This test is similar to test 3. However, in this test the rat is positioned so that the side of its body is moved towards the edge of the counter top. In normal animals both forelimb and hindlimb are extended to touch the edge of the counter top. Animals with unilateral damage are unable to extend the contralateral forelimb and hindlimb.

Test 5: The rat is placed on a counter top and pushed nose first towards the edge. Normal rats will place both forepaws beneath its body to prevent themselves from going over the edge. In contrast, rats with unilateral hemorrhagic injury are unable to use their contralateral forelimb, which drops over the edge of the bench top.

Test 6: This test is similar to test 5, but the animal is pushed sideways towards the edge of the counter top. A normal rat will use both its forelimb and hindlimb to hold the edge by placing them beneath their body. Rats with unilateral hemorrhages

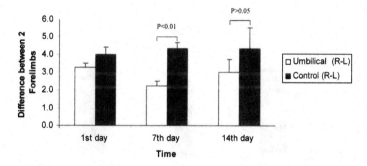

FIGURE 1. Effects of human cord blood infusion on stepping performance in hemorrhagic rats. The stepping test measures the difference in the number of steps taken by an animal between its left and right forelimbs over a fixed distance (1 meter). The difference between the left and right forelimbs is typically zero. Hemorrhagic rats treated with cord blood displayed significant reductions in stepping performance compared to untreated controls at day 7 after hemorrhagic brain injury ($P < 0.05$).

are unable to use their contralateral forelimbs and hindlimbs in this fashion and these limbs droop over the edge of the counter top.

Stepping Test

In the stepping test animals are gently held so that only one forelimb extends freely from its body. The animal is then positioned so that the forelimb comes into contact with a bench top. As the suspended animal is moved forward along the bench top, its forelimb will reflexly make a stepping movement. The number of stepping movements made by each animal is quantified over a distance of 100 cm for both the left and right forelimb. In normal animals the difference in stepping movements between the left and right forelimb is insignificant. Animals with hemorrhagic stroke to one side of the brain, however, exhibit significantly fewer steps by the contralaterally affected forelimb in comparison to the nonaffected ipsilateral forelimb.

Elevated Body-Swing Test

In the elevated body-swing test (EBST), animals are held by the tail, and suspended vertically. While in this vertical position, the animals are monitored to determine whether they swing or turn to the left or right in order to reach upward. This process is repeated on 10 separate occasions. Normal animals will reach upwards to the left and right an equal number of times and exhibit no bias to one particular side. Consequently, the bias difference should be zero. In contrast, animals with unilateral hemorrhagic brain injury generally exhibit a dramatic bias to one side.

Immunohistochemical Analysis

We use diaminobenzidine (DAB) staining to look for human-derived umbilical cord blood cells in rat brain, and fluorescent double-staining to determine cell phenotype. For DAB staining, free-floating sections were processed by standard procedures. After quenching and a PBS rinse, 5% normal horse serum plus 0.3% Triton in PBS was used for blocking and subsequent antibody incubations. A mouse anti-

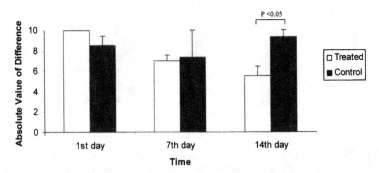

FIGURE 2. Effects of human cord blood infusion on elevated body-swing test in hemorrhagic rats. Animals treated with cord blood exhibited a progressive decrease in elevated body-swing bias over the 14 days of testing. By day 14 there was a significant decrease in elevated body-swing bias between cord blood–treated animals and untreated controls ($P < 0.05$).

human nuclei antibody (1:200, Chemicon International, Temecula, CA; MAB 1281) was used as the primary antibody for overnight incubation at 4°C, and a biotinylated horse anti-mouse antibody (1:200, Vector Laboratories Inc, Burlingame, CA) was used as secondary antibody for a 1-hour incubation at room temperature. This was followed by a 1-hour incubation in an ABC solution (Vector Laboratories Inc, Burlingame, CA; Vectastain PK-4000 ABC kit). After rinsing in PBS, the DAB solution (Vector; DAB kit SK-4100) was added for 5 minutes. Sections were rinsed again in PBS, mounted, dehydrated in a graded ethanol series, and coverslipped. For fluorescent staining, free-floating sections were blocked with 5% normal donkey serum plus 0.3% Triton (1 hour) in PBS, followed by incubation in mouse anti-human nuclei (1:200, Chemicon International, Temecula, CA; MAB 1281) overnight at 4°C. After rinsing in PBS, Cy3-conjugated AffiniPure donkey anti-mouse (1:200, Jackson ImmunoResearch; 62691) was added for a 2-hour incubation at room temperature. After rinsing, a second primary antibody, either rabbit anti-GFAP (1:1500, DAKOcytomation; Z 0334) or mouse anti-NeuN (1:200, Chemicon International, Temecula, CA; MAB 377) was added for another 2-hour incubation at room temperature. This was followed by an incubation in Alexa Fluor 488 donkey anti-rabbit or donkey anti-mouse (1:400, Molecular Probes, Eugene, OR) for 1 hour at room temperature. Sections were rinsed again, mounted, and coverslipped. A Nikon Eclipse E-600 microscope equipped with SPOT INSIGHT camera and software from Diagnostic Instruments was used for analyzing cell type and differentiation from human umbilical cord blood cells.

Statistical Analyses

Statistical comparisons of the behavioral data among the different experimental groups were made using analysis of variance (ANOVA). Pair-wise post hoc comparisons were made among experimental groups when F values for ANOVA indicate significant differences. P values < 0.05 were considered significant.

RESULTS

In the stepping test animals were supported by the torso, and either the left or right forelimb was placed in contact with a counter top. As the animal was moved forward, a reflex stepping response was initiated. In normal rats the number of steps taken over a fixed distance is the same for the left and right forelimb, and thus the difference in the number of steps is zero. In animals with unilateral hemorrhagic brain injury, however, the affected contralateral limb takes fewer steps, and thus the difference between the two forelimbs is greater than zero. In our examination of the stepping response, the difference in the number of steps taken by the left and right forelimbs was significantly less for rats treated with UCB than for untreated rats at the 7- and 14-day time periods (Fig. 1). At day 1 prior to the administration of cord blood, there was no difference between the UCB and the untreated animals: 3.3 ± 0.5 vs. 4.0 ± 0.4 (mean $\pm$ SEM), respectively ($P > 0.05$). At day 7 the differences were 2.3 ± 0.3 vs. 4.3 ± 0.3 ($P < 0.05$) for UCB and untreated animals, respectively. At day 14 the differences were 3.0 ± 0.7 vs. 4.3 ± 1.2 ($P < 0.05$) for UCB and untreated animals, respectively.

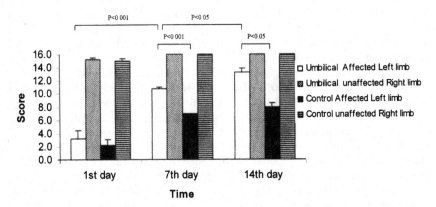

FIGURE 3. Effects of human cord blood infusion on limb-placement function in hemorrhagic rats. Limb-placement test measures combined sensorimotor deficits. Animals treated with cord blood showed progressive improvement in limb-placement function over the 14 days of testing. Cord blood–treated animals exhibited significantly better limb-placement performance compared to control animals at 7 and 14 days after hemorrhagic brain injury ($P < 0.05$).

In the elevated body-swing test, animals were suspended by the tail and monitored as they reached up to either the right or left. When this process is done repeatedly, normal animals typically show very little bias in reaching up to the right or left. In contrast, animals with unilateral hemorrhagic brain injury g enerally exhibit a dramatic bias. When we evaluated rats at day 1 prior to cord blood administration, we found that EBST bias was not significantly different between UCB and untreated animals (10 ± 0.0 vs. 8.5 ± 0.9, respectively). At day 7 after cord blood administration there was still no significant difference between these two groups (7.0 ± 0.6 vs. 7.5 ± 2.7, respectively). However, by day 14 rats treated with UCB exhibited a significant decrease in EBST bias in comparison to non-treated controls (5.5 ± 0.9 vs. 9.3 ± 0.7, $P < 0.05$), respectively (FIG. 2).

In the limb placement test, animals were examined for their ability to use the left and right forelimbs and hindlimbs. When we evaluated rats at day 1 prior to cord blood administration, UCB and untreated rats were not significantly different (3.3 ± 1.2 vs. 2.3 ± 0.9, respectively). By day 7, however, there was a significant improvement in UCB-treated rats in comparison to the untreated controls (10.8 ± 0.3 vs. 7.0 ± 0.0, $P < 0.05$). This improvement was further accentuated by day 14 (13.3 ± 0.6 vs. 8.0 ± 0.6, $P < 0.05$). At this time point, the treated animals were near normal in their placement responses (FIG. 3).

Histologic analysis for human-derived cells revealed MAB 1281–positive cells primarily in the area of the penumbra surrounding the hemorrhagic injury. Scattered cells were also observed in the striatum and cortex in the lesioned side of the brain. Scant cells were seen in the contralateral hemisphere. In general, the number of MAB 1281–positive cells in animals with infused cord blood were surprisingly few. Double-labeling was performed to determine the phenotype of the infused cells. Im-

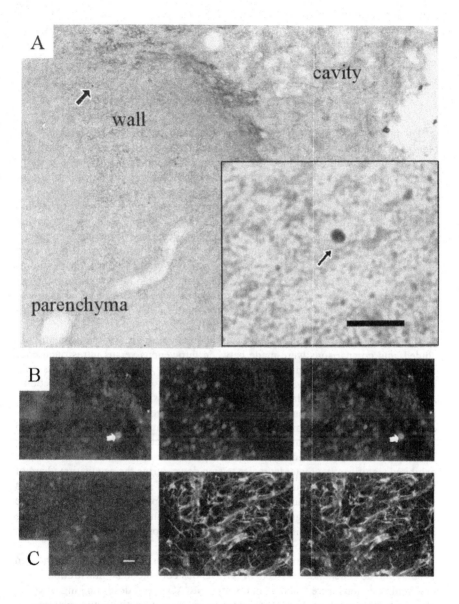

FIGURE 4. Histologic analysis of infused human cord blood. (**A**) DAB immunostaining for human nuclei with MAB 1281staining. UCB cell (*arrows* in both lower- and higher-magnification panel) is shown in the wall of cavity formed from hemorrhage. *Bar* in high-magnification panel equals 25 μm. (**B**) Fluorescent staining for human nuclei–positive cell found in the wall of the lesion cavity. *Arrow* in left panel indicates MAB 1281–positive cell. Middle panel shows staining for NeuN. Right panel is merged image showing lack of co-localization of MAB 1281 and NeuN. (**C**) Left panel shows fluorescent staining for MAB 1281. Middle panel shows staining for GFAP. Right panel is merged image showing co-localization of MAB 1281 and GFAP.

munofluorescent histochemistry for MAB 1281 and NeuN revealed no MAB 1281 cells that were also positive for NeuN (FIG. 4). However, MAB 1281 and GFAP double-labeled cells were observed.

DISCUSSION

The results from this study demonstrate that infusions of human umbilical cord blood can ameliorate neurologic deficits associated with intracerebral hemorrhage. Animals treated with cord blood exhibited improvements in a neurologic severity test, a stepping test, and an elevated body-swing test. These results are consistent with other studies that have utilized cord blood to treat ischemic brain injury.[8,18,19] In the study by Chen et al.[8] transient ischemic injury was induced by occluding the middle cerebral artery for 2 hours. Human cord blood cells (3×10^6) were injected into the tail vein at 1 or 7 days after MCAO. Significant improvements were found in rotarod performance and a modified neurologic severity test in animals given cord blood at 1 day after MCAO when evaluated over a period of 14 days. These studies were the first to demonstrate the efficacy of human cord blood administration in experimental stroke, and showed that early treatment was more efficacious than late treatment. In these studies, cord blood administration at day 7 after MCAO was significantly less effective than treatment at day 1.

In the study by Willing et al.[18] the middle cerebral artery was permanently occluded, and comparisons were made between intravenous versus intrastriatal cord blood administration. For intravenous administration, 1×10^6 cord blood cells were injected into the femoral vein 24 hours after MCAO. For intrastriatal injections, 2.5×10^5 cells were transplanted intraparenchymally using stereotactic techniques, also 24 hours after occlusion of the middle cerebral artery. In measures of spontaneous activity evaluated at 1 and 2 months after transplantation, both intravenous and intrastriatal cord blood groups showed significant reductions in hyperactive activity in comparison to non-treated animals. In the passive avoidance test, treated animals exhibited significantly greater latencies to step down during the acquisition phase of this test in comparison to non-treated controls. In the stepping test, only animals treated with intravenous administration exhibited improved performance. These studies by Willing et al. suggest that intravenous and intrastriatal administration are somewhat equivalent in their efficacy except for the stepping test where intravenous administration was found to be superior.

In the studies by Vendrame et al.,[19] permanent occlusions of the MCA were used to study the dose-dependent effects of intravenous cord blood administration. Femoral vein injections were made 24 hours after MCAO at cell doses ranging from 1×10^4 to 5×10^7 cells. Significant reductions in spontaneous activity were observed in animals treated with low and high doses of cord blood cells in comparison to media-treated controls. Reductions in elevated body-swing bias were observed only in the high-dose groups ($\geq 10^6$ cells). Likewise, improvements in the stepping test were also seen only in the high-dose groups ($\geq 10^5$ cells). These results found by Vendrame et al. pinpoint cell doses required for therapeutic benefit by cord blood administration in the rodent with ischemic brain injury, and appear to be comparable for hemorrhagic brain injury.

Other similar types of cells have also been found to provide therapeutic benefit after transplantation in ischemic brain injury. The efficacy of bone marrow cells, for example, has been studied in rodents with MCA occlusions. In the study by Chen *et al.*[8] rodent bone marrow was administered intravenously at cell doses of 1×10^6 and 3×10^6 cells one day after transient MCA occlusion. Rats that received the higher dose of cells exhibited functional improvement in an adhesive removal test, a rotarod test, and in a neurologic severity score test. Intra-arterial administration of 2×10^6 bone marrow cells one day after transient MCA occlusion also has been demonstrated to provide therapeutic benefit,[20] as did intracerebral[8] and intrastriatal[21] transplants. Although the results of these cord blood and bone marrow studies demonstrate functional therapeutic benefits, the type(s) of cells that mediate these restorative effects have yet to be identified.

Recently, non-hematopoietic progenitor cells have been isolated from bone marrow[22] and demonstrated to differentiate into endoderm, mesoderm, and ectoderm.[23] These multipotent adult progenitor cells (MAPCs) have been transplanted into to the cortex of rats with permanent occlusions of the distal aspect of the middle cerebral artery.[24] This type of occlusion results in the selective damage to the sensorimotor cortex of rats. Rats with cortical transplants of human MAPCs (2.25×10^5 cells) implanted 7 days after cerebral ischemic injury displayed significant recovery of limb-placement function as well as enhanced performance in a sensorimotor adhesive tape–removal test. These MAPCs may represent the cells within bone marrow that provide functional benefit seen in animals with ischemic brain injury.

Neural progenitor cells and neural stem cells have also demonstrated therapeutic benefits in ischemic or hemorrhagic brain injury. Jansen *et al.*[25] transplanted cortical progenitor cells (10^5 cells) from fetal rats into the cortex of postnatal day 10 rats three days after unilateral hypoxic-ischemic brain injury. Treated animals exhibited improved performance on a rotarod and for apomorphine-induced rotational asymmetry in comparison to hypoxic-ischemic control animals. Onifer and Low[26] transplanted neural progenitor cells from the hippocampal formation of fetal gerbils into adult gerbils with bilateral transient occlusion of the common carotid arteries. This type of lesion resulted in the destruction of pyramidal cells in the CA1 region of the hippocampus leading to spatial memory deficits. Bilateral transplants of 7×10^4 cells into the damaged hippocampus 7–10 weeks after the ischemic injury were able to restore spatial memory function as assessed using a circular water maze. Similar therapeutic effects were observed using conditionally immortalized hippocampal neuroepithelial cells.[27] Transplants of 5×10^4 cells into the CA1 region of rats 2–3 weeks after global ischemia induced by the transient 4-vessel occlusion method resulted in improved spatial navigation performance. Together these results suggest that neural progenitor cells from rodents can provide therapeutic benefits in ischemic brain injury.

The use of human neural stem cells can also provide therapeutic benefit in ischemic brain injury. Neural stem cells derived from human fetal forebrain at a gestational age of 7 weeks[28] were transplanted at a dose of 5×10^5 cells into the caudate nucleus 4 days after the common carotid artery was transiently occluded in gerbils. Animals with neural stem cell grafts exhibited significantly better performance in an elevated body swing test, an asymmetry test, and a T-maze test in comparison to sham-treated control animals. The effects of intravenous infusions of human neural stem cells in experimental cerebral ischemia have been studied by Chu *et al.*[29] Neural stem cells from human fetal cortical tissue 15 weeks in gestation were infused into the tail vein of rats (5×10^6 cells) one day after transient MCA occlusion. Animals infused with human neural stem cells exhibited enhanced performance in a rotarod test, a modified limb-placement test, and a turning alley test. These results suggest that human neural stem cells can provide therapeutic benefits in experimental ischemic brain injury after both intraparenchymal injections and systemic infusions.

Neural stem cells also provide therapeutic benefit in experimental hemorrhagic brain injury. A study by Jeong et al.[30] administered human neural stem cells intravenously in rats with collagenase-induced intracerebral hemorrhage. In this study, 5×10^6 cells derived from human fetal cortical tissue 15 weeks in gestation were infused by the tail vein one day after hemorrhagic brain injury. Treated animals evaluated in rotarod and limb-placement tests showed significantly better performances in comparison to saline-treated controls over an 8-week testing period.

From this review of the literature it is clear that several types of cells can provide therapeutic benefit for treating hemorrhagic and ischemic brain injury. A general theme that emerges is that intravascular infusions of cells within 1 day after injury can provide therapeutic benefit, while intracerebral transplants can be made as late as 10 weeks after injury and still enhance functional recovery. Thus it may be possible to devise different cell therapy strategies for treating acute vs. chronic ischemic or hemorrhagic brain injury.

One of the surprising results from the present study was the paucity of human cord blood cells found within the parenchyma of the brain around the site of injury in spite of the functional recovery observed in the treated animals. Similar observations have been made by Vendrame et al.[19] in their study using cord blood. In this study they found scant cord blood cells in the brain on the side of the injury even in animals that had received high doses of cord blood cells that produced significant sparing of host brain tissue. A study by Borlongan et al.[31] suggests that cord blood cells need not penetrate into the brain parenchyma to induce therapeutic effects. These results suggest that mechanisms other than cell replacement may underlie the restorative effects of cord blood cells.

Studies have demonstrated that cord blood contains growth factors similar to those that exert neurotrophic effects. Neurotrophins such as brain-derived neurotrophic factor (BDNF), neurotrophin-3 (NT3), and nerve growth factor (NGF) have been found in cord blood and and cord blood serum.[32-36] These neurotrophins are known to have profound effects on brain plasticity.[37,38] In particular, the studies of BDNF suggest that this neurotrophin plays a critical role in the development and alterations in axodendritic connections. Injections of BDNF have been shown to extend critical periods of development for climbing fiber plasticity in the cerebellum of rats,[37] and enhance nerve fiber outgrowth in the arcopallium of adult song birds to induce song sequence variability that is typically seen in juvenile birds.[38] Thus the restorative effects observed in the present study with cord blood may have been mediated in part by the secretion of neurotrophins.

In summary we have shown that intravenous infusions of cord blood into rats with intracerebral hemorrhage can ameliorate neurologic deficits associated with hemorrhagic brain injury. These results suggest that the administration of human cord blood may provide a therapeutic approach for treating patients who have experienced hypertension-induced bleeding in, or trauma or penetrating wound injuries to, the brain.

ACKNOWLEDGMENTS

This work was supported in part by funds from the Lyle French Endowment, and PHS Grant NIH RO1-40831. This work was done in collaboration with Saneron CCEL Therapeutics, Inc. P.R.S. is a founder of that company and W.C.L. is a consultant. We thank Jordan Webb for administrative assistance.

REFERENCES

1. INAGAWA, T. 2002. What are the actual incidence and mortality rates of intracerebral hemorrhage? Neurosurg. Rev. **25:** 237–246.
2. MOHR, J.P., L.R. CAPLAN, J.W. MELSKI, *et al.* 1978. The Harvard Cooperative Stroke Registry: a prospective registry. Neurology **28:** 754–762.
3. BOGOUSSLAVSKY, J., G. VAN MELLE & F. REGLI. 1988. The Lausanne Stroke Registry. Stroke **19:** 1083.
4. FOULKES, M.A., P.A. WOLF, T.R. PRICE, *et al.* 1988. The Stroke Data Bank: design, methods and baseline characteristics. Stroke **19:** 547–554.
5. FISHER, C.M. 1961. Clinical syndromes in cerebral hemorrhage. *In* Pathogenesis and Treatment of Cerebrovascular Disease. W.S. Fields, Ed. Charles C Thomas. Springfield, IL.
6. HIER, D.B., K.R. DAVIS, E.P. RICHARDSON, JR. & J.P. MOHR. 1977. Hypertensive putaminal hemorrhage. Ann. Neurol. **1:** 152–159.
7. CAPLAN, L.R. & J.P. MOHR. 1978. Intracerebral hemorrhage. Geriatrics **33:** 42.
8. CHEN, J., P.R. SANBERG, Y. LI, *et al.* 2001. Intravenous administration of human umbilical cord blood reduces behavioral deficits after stroke in rats. Stroke **32:** 2682–2688.
9. LU, D., P.R. SANBERG, A. MAHMOOD, *et al.* 2002. Intravenous administration of human umbilical cord blood reduces neurological deficit in the rat after traumatic brain injury. Cell Transplant. **11:** 275–281.
10. BROXMEYER, H.E., G.W. DOUGLAS, G. HANGOC, *et al.* 1989. Human umbilical cord blood as a potential source of transplantable hematopoietic stem/progenitor cells. Proc. Natl. Acad. Sci. USA **86:** 3828–3832.
11. ERICES, A., P. CONGET & J.J. MINGUELL. 2000. Mesenchymal progenitor cells in human umbilical cord blood. Br. J. Haematol. **109:** 235–242.
12. MUROHARA, T., H. IKEDA, J. DUAN, *et al.* 2000. Transplanted cord blood-derived endothelial precursor cells augment postnatal neovascularization. J. Clin. Invest. **105:** 1527–1536.
13. ROSENBERG, G.A., S. MUN-BRYCE, M. WESLEY & M. KORNFELD. 1990. Collagenase-induced intracerebral hemorrhage in rats. Stroke **21:** 801–807.
14. JANOFF, A. & J.D. ZELIGS. 1968. Vascular injury and lysis of basement membrane in vitro by neutral protease of human leukocytes. Science **161:** 702–704.
15. WEISS, S.J. 1989. Tissue destruction by neurophilis. N. Engl. J. Med. **320:** 365–376.
16. CHESNEY, J.A., T. KONDOH, J.A. CONRAD & W.C. LOW. 1995. Collagenase-induced intrastriatal hemorrhage in rats results in long-term locomotor deficits. Stroke **26:** 312–316.
17. DE RYCK, M. 1990. Animal models of cerebral stroke: pharmacological protection of function. Eur. Neurol. (Suppl. 2) **30:** 21–27; discussion, 39–41.
18. WILLING, A.E., J. LIXIAN, M. MILLIKEN, *et al.* 2003. Intravenous versus intrastriatal cord blood administration in a rodent model of stroke. J. Neurosci. Res. **73:** 296–307.
19. VENDRAME, M., J. CASSADY, J. NEWCOMB, *et al.* 2004. Infusion of human umbilical cord blood cells in a rat model of stroke dose-dependently rescues behavioral deficits and reduces infarct volume. Stroke **35:** 2390–2395.
20. LI, Y., J. CHEN, L. WANG, *et al.* 2001. Treatment of stroke in rat with intracarotid administration of marrow stromal cells. Neurol. **56:** 1666–1672.

21. Li, Y. M. Chopp. J. Chen, *et al.* 2001. Intrastriatal transplantation of bone marrow non-hematopoietic cells improves functional recovery after stroke in adult mice. J. Cereb. Blood Flow and Metab. **20:** 1311–1319.

22. Reyes, M., T. Lund, T. Lenvik, *et al.* 2001. Purification and ex vivo expansion of postnatal human marrow mesodermal progenitor cells. Blood **98:** 2615–2625.

23. Jiang, Y., B.N. Jahagirdar, R.L. Reinhardt, *et al.* 2002. Pluripotency of mesenchymal stem cells derived from adult marrow. Nature **418:** 41–49.

24. Zhao, L.R., W.M. Duan, M. Reyes, *et al.* 2002. Human bone marrow stem cells exhibit neural phenotypes and ameliorate neurological deficits after grafting into the ischemic brain of rats. Exp. Neurol. **174:** 11–20.

25. Jansen E.M. L. Solberg, S. Underhill, *et al.* 1997. Transplantation of fetal neocortex ameliorates sensorimotor and locomotor deficits following neonatal ischemic-hypoxic brain injury in rats. Exp. Neurol. **147:** 487–497.

26. Onifer, S.M. & W.C. Low. 1990. Spatial memory deficit resulting from ischemia-induced damage to the hippocampus is ameliorated by intra-hippocampal transplants of fetal hippocampal neurons. Progr. Brain Res. **82:** 359–366.

27. Sinden, J.D., F. Rashid-Doubell, T.R. Kershaw, *et al.* 1997. Recovery of spatial learning by grafts of a conditionally immortalized hippocampal neuroepithelial cell line into the ischaemia-lesioned hippocampus. Neuroscience **81:** 599–608.

28. Ishibashi, S., M. Sakaguchi, T. Kuroiwa, *et al.* 2004. Human neural stem/progenitor cells, expanded in long-term neurosphere culture, promote functional recovery after focal ischemia in Mongolian gerbils. J. Neurosci. Res. **78:** 215–223.

29. Chu, K., M. Kim, K.I. Park, *et al.* 2004. Human neural stem cells improve sensorimotor deficits in the adult rat brain with experimental focal ischemia. Brain Res. **1016:** 145–153

30. Jeong, S.W., K. Chu, K.H. Jung, *et al.* 2003. Human neural stem cell transplantation promotes functional recovery in rats with experimental intracerebral hemorrhage. Stroke **34:** 2258–2263.

31. Borlongan, C.V., M. Hadman, C.D. Sanberg & P.R. Sanberg. 2004. Central nervous system entry of peripherally injected umbilical cord blood cells is not required for neuroprotection in stroke. Stroke **35:** 2385–2389.

32. Walker, P., R.H. Tarris, M.E. Weichsel, Jr., *et al.* 1981. Nerve growth factor in human umbilical cord serum: demonstration of a veno-arterial gradient. J. Clin. Endocrinol. Metab. **53:** 218–220.

33. Haddad, J., V. Vilge, J.G. Juif, *et al.* 1994. Beta-nerve growth factor levels in newborn cord sera. Pediatr. Res. **35:** 637–639.

34. Simone, M.D., S. De Santis, E. Vigneti, *et al.*. 1999. Nerve growth factor: a survey of activity on immune and hematopoietic cells. Hematol. Oncol. **17:** 1–10.

35. Bracci-Laudiero, L., D. Celestino, G. Starace, *et al.* 2003. CD34-positive cells in human umbilical cord blood express nerve growth factor and its specific receptor TrkA. J. Neuroimmunol. **136:** 130–139.

36. Chouthai, N.S., J. Sampers, N. Desai & G.M. Smith. 2003. Changes in neurotrophin levels in umbilical cord blood from infants with different gestational ages and clinical conditions. Pediatr. Res. **53:** 965–969.

37. Sherrad, R.M. & A.J. Bower. 2001. BDNF and NT3 extend the critical period for developmental climbing fiber plasticity. NeuroReport. **12:** 2871–2874.

38. Kittleberger, J.M. & R. Mooney. 2005. Acute injections of brain-derived neurotrophic factor in a vocal premotor nucleus reversibly disrupt adult birdsong stability and trigger syllable deletion. J. Neurobiol. **62:** 406–424. [Published online November 16, 2004; <www.interscience.wiley.com/DOI 10.1002/neu.20209>.]

How Wnt Signaling Affects Bone Repair by Mesenchymal Stem Cells from the Bone Marrow

CARL A. GREGORY, WILLIAM G. GUNN, EMIGDIO REYES,
ANGELA J. SMOLARZ, JAMES MUNOZ, JEFFREY L. SPEES,
AND DARWIN J. PROCKOP

*Center for Gene Therapy, Tulane University Health Sciences Center,
New Orleans. Louisiana 70112, USA*

ABSTRACT: Human mesenchymal stem cells (hMSCs) from bone marrow are a source of osteoblast progenitors *in vivo*, and under appropriate conditions they differentiate into osteoblasts *ex vivo*. The cells provide a convenient cell culture model for the study of osteogenic tissue repair in an experimentally accessible system. Recent advances in the field of skeletal development and osteogenesis have demonstrated that signaling through the canonical wingless (Wnt) pathway is critical for the differentiation of progenitor cell lines into osteoblasts. Inhibition of such signals can predispose hMSCs to cell cycle entry and prevent osteogenesis. Our investigation of the role of Wnt signaling in osteogenesis by hMSCs *ex vivo* has demonstrated that osteogenesis proceeds in response to bone morphogenic protein 2 stimulation and is sustained by Wnt signaling. In the presence of Dkk-1, an inhibitor of Wnt signaling, the cascade is disrupted, resulting in inhibition of osteogenesis. Peptide mapping studies have provided peptide Dkk-1 agonists and the opportunity for the production of blocking antibodies. Anti-Dkk-1 strategies are clinically relevant since high serum levels of Dkk-1 are thought to contribute to osteolytic lesion formation in multiple myeloma and possibly some forms of osteosarcoma. Specific inhibitors of glycogen synthetase kinase 3β (GSK3β), which mimic Wnt signaling, may also have a therapeutic benefit by enhancing *in vitro* osteogenesis despite the presence of Dkk-1. Antibodies that block Dkk-1 and GSK3β inhibitors may provide novel opportunities for the enhancement of bone repair in a variety of human diseases such as multiple myeloma and osteosarcoma.

KEYWORDS: mesenchymal stem cells; Wnt; Dkk-1; bone; osteogenesis; cancer

MESENCHYMAL STEM CELLS FROM BONE MARROW

The first non-hematopoietic mesenchymal stem cells (MSCs) were discovered by Friedenstein,[1] who described clonal, plastic adherent cells from bone marrow capable of differentiating into osteoblasts, adipocytes, and chondrocytes.[2–5] These cells

Address for correspondence: Carl A. Gregory, Center for Gene Therapy, Tulane University Health Sciences Center, 1430 Tulane Avenue, New Orleans, LA 70112. Voice: 504-988-7716; fax: 504-988-7710.
ca_gregory@hotmail.com

Ann. N.Y. Acad. Sci. 1049: 97–106 (2005). © 2005 New York Academy of Sciences.
doi: 10.1196/annals.1334.010

were later found to differentiate into "stromal" cells, structural components of the bone marrow that supported *ex vivo* culture of hematopoiesis by providing extracellular matrix components, cytokines, and growth factors.[6–10] Numerous laboratories around the world have now demonstrated that multipotent MSCs can be recovered from a variety of adult tissues and can differentiate into various tissue lineages. In particular, Verfaille and colleagues report that a specific type of murine MSC isolated from bone marrow, muscle or brain termed multipotential adult progenitor cells (MAPCs) differentiate into a variety of tissue lineages including myoblasts, hepatocytes, and even neural tissue.[11–14]

Although it is clear that single cell–derived colonies of MSCs can transdifferentiate into multiple tissue lineages in culture, there have been indications that they can also undergo cell fusion. Evidence supporting the case for cell fusion by MSCs comes from a set of experiments conducted by Spees and colleagues[15] in which MSCs, labeled with green fluorescent protein, were co-cultured with pulmonary small airway epithelial cells (SAECs) after a brief heat shock to mimic tissue damage. After heat shock, a good proportion of the confluent SAEC monolayer died by apoptosis. When introduced to the damaged SAEC cultures, the green MSCs adhered to the tissue culture plastic, resulting in a confluent mixture of cells. Over a few days, some of the MSCs differentiated into epithelial-like cells but others fused with SAECs, producing chimeras. These investigators suggested that cell fusion may be an acute response to tissue damage and may rescue dying cells by supplying additional cytosolic metabolites and organellar components.

Because multipotent MSCs are easily expanded in culture, there has been much interest in their clinical potential for tissue repair and gene therapy.[16] As a result, numerous studies have been carried out demonstrating the migration and multi-organ engraftment potential of MSCs in animal models and in human clinical trials.[13,17–25] One trial in particular utilized hMSCs from compatible healthy donors to treat individuals with the brittle bone disease osteogenesis imperfecta, yielding encouraging results.[18–20]

In addition to their ability to home, engraft, and differentiate into damaged tissues, MSCs secrete cytokines and trophic factors that provide beneficial effects to surrounding tissue. This characteristic has been exploited in the field of regenerative neuroscience, where MSCs, when injected into the brain or damaged spinal cord, engraft and provide a permissive stromal microenvironment for the growth and repair of existing neural tissue.[22,26–29] In our laboratory, MSCs have been shown to express and secrete a variety of neurotrophic and neuroprotective proteins in culture and when implanted into the brains of test animals (Munoz *et al.*, unpublished material). Expression of these beneficial factors persists from stably engrafted MSCs for a number of weeks post implantation, and it is hoped that this discovery will lead to a promising treatment for neural lesions caused by ischemia and spinal cord injury. Although the MSCs secrete their own extensive repertoire of cytokines, MSCs can be genetically modified in a stable manner by a variety of retroviral and lentiviral vectors and therefore could be induced to secrete agents at sites of tissue injury or tumors. It is noteworthy that MSCs have been shown to preferentially migrate to tumors and, when genetically modified to secrete interferon-β, they reduce the size of the lesions.[36]

Graft and host compatibility is always an issue when transplanting heterologous tissue into an immunocompetent host, and immunotolerance of implanted MSCs has been the subject of much debate in recent years. One remarkable facet of MSC physiology is that the cells may actually inhibit inflammation and immunologic responses in the host. *In vitro,* MSCs fail to induce allogeneic responses in mixed lymphocyte reaction assays and they escape lysis by cytotoxic T cells and natural killer cells. The immunomodulatory properties of MSCs are probably explained by their lack of an HLA type II receptor and the secretion of cytokines.[30–33] Indeed, allogenic MSC implants do not appear immunogenic in the non-human primate[34] and human recipients receiving MSCs from sibling donors.[18,19] However, the requirement for fetal bovine serum (FBS) in the expansion medium of MSCs is a source for concern when administering the cells to human recipients. Using an assay based on fluorescently labeled FBS it was found that a dose of 10 million MSCs could be contaminated with milligram quantities of FBS. Furthermore, most of the FBS was internalized and could not be removed by washing the cells.[35] This observation explains the findings of Horwitz *et al.,* who reported that after an infusion of allogenic MSCs, 5 of 6 children with ostoegenesis imperfecta showed improved growth and skeletal durability, whereas one recipient showed no improvement (patient 6 in the study) and reacted immunologically to the infusion of cells.[20] Testing revealed that the patient had developed a profound humoral response against bovine serum albumin, the major bovine contaminant in MSC preparations. The solution to this problem lies in the ability of MSCs to expand in adult human serum and retain all of their beneficial properties. In the clinic, MSCs could be expanded in FBS to the desired number, then grown for an additional 2 days in medium containing the patients' own serum. Support for this strategy is based on findings from a rat model, where repeated administrations of rat MSCs evoked a profound humoral response against FBS. However, when expanded for a short time in medium containing rat serum, the response did not occur.[35] Furthermore, in human cultures of MSCs, FBS contamination could be reduced to nearly zero by a 2-day culture in medium containing human serum[35] and should therefore prevent the occurrence of immune reactions in subsequent trials.

MODULATION OF WNT SIGNALING AFFECTS GROWTH AND OSTEOGENIC DIFFERENTIATION OF MSCs

Since human MSCs (hMSCs) are recovered from a donor by a straightforward iliac crest aspirate and simply enriched by virtue of their adherence to tissue culture plastic, the cells provide a convenient *ex vivo* model for the study of mesenchymal stem cell expansion and differentiation. Cultures of hMSCs classically display a lag phase of about 4 days before they enter a phase of exponential growth.[37–39] On further investigation, it was found that conditioned medium from cultures of hMSCs reduced the lag period when added to freshly plated MSCs[39] and ablated a shorter 12-hour lag period when the medium of established hMSC cultures was replaced[10] (see FIG. 1a). The agent responsible for the lag-phase ablation was identified as Dickkopf-1 (Dkk-1), an inhibitor of the canonical Wnt pathway.[10,40,41] Dkk-1 was highly expressed in rapidly dividing hMSCs, but was absent or expressed at very low levels at the stationary phase of growth and when proliferation was blocked by withdrawal of serum (FIGS. 1b, c, and d). Recombinant Dkk-1 reduced the length of a lag phase induced by a change of medium (FIG. 1e) and addition of antiserum against Dkk-1

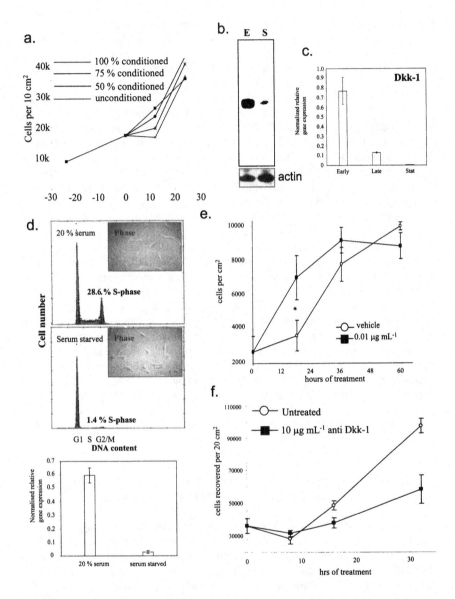

FIGURE 1. (**a**) Growth of hMSCs after medium replacement containing various proportions of conditioned medium. (**b**) Western blot assay for Dkk-1 in hMSCs from early log (E) and stationary (S) cultures of MSCs. (**c**) Hybridization ELISA analysis of Dkk-1 encoding RT-PCR products generated from early, late, and stationary phase cultures. Dkk-1 expression is high during proliferation. Signals were normalized against GAPDH expression. (**d**, *upper 2 panels*) During growth arrest by serum starvation, transcription of Dkk-1 is inhibited. Cell cycle analysis of hMSCs after 5 days in culture followed by addition of medium containing no FCS or 20 % (v/v) FCS. The relative proportions of cells in G1, S phase, and G2 phase are indicated on the histograms. Phase-contrast micrographs are presented with each histogram illustrating cell density in each case.(**d**, *lower panel*) Hybridization ELISA

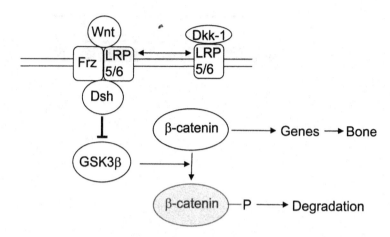

FIGURE 2. Simplified scheme of the canonical Wnt pathway. Explanation of the abbreviations are given in the text.

extended the lag phase (FIG. 1f). Interestingly, these observations seem consistent with the model for Wnt signaling during limb bud development proposed by Hartmann and Tabin,[42] where positive signaling through the canonical Wnt pathway induces differentiation and a postmitotic state by increasing the intracellular level of stable β-catenin.

In the canonical pathway (FIG. 2), Wnt ligands bind to the transmembrane receptor frizzled (Frz) and the co-receptor lipoprotein-related protein 5 and 6 (LRP-5/6). Activation of frizzled recruits the cytoplasmic bridging molecule, disheveled (Dsh), so as to inhibit glycogen synthetase kinase 3β (GSK3β). Inhibition of GSK3β decreases phosphorylation of β-catenin, preventing its degradation by the ubiquitin-mediated pathway. The stabilized β-catenin acts on the nucleus by activating TCF/LEF-mediated transcription of target genes.[43,44] In the light of recent work demonstrating that canonical Wnt signaling drives the differentiation of progenitor cell lines into osteoblasts,[45,46] it seems reasonable that Dkk-1 is acting to prevent terminal differentiation of the MSCs in culture while they proliferate. Two observations support this hypothesis: (i) when Wnt signaling is mimicked by culturing the MSCs with the glycogen synthetase kinase-β inhibitor, LiCl, osteogenesis can be accelerated (FIGS. 3a and b), and (ii) addition of exogenous Dkk-1 can inhibit osteogenesis (FIG. 3c).

---

analysis of Dkk-1 encoding PCR products demonstrating Dkk-1 is only expressed during proliferation. (**e**) Effect of Dkk-1 on the lag phase of hMSCs. Conditioned medium of lag-phase cells was replaced with fresh medium containing vehicle or 0.01 μg mL^{-1} recombinant Dkk-1. (**f**) Effect of anti-Dkk-1 polyclonal serum on proliferation of hMSCs after a change of medium to induce a second lag period. (Data reproduced with permission from Gregory *et al.*[10])

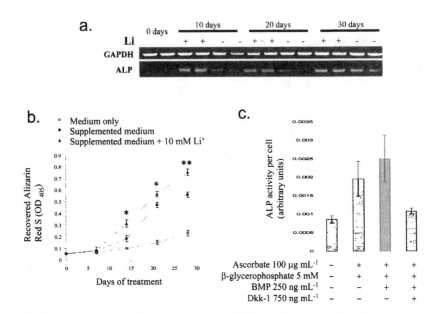

FIGURE 3. Long-term treatment of hMSCs with 10 mM LiCl improves the differentiation of hMSCs into osteoblasts. (**a**) RT-PCR assays for alkaline phosphatase (ALP), a marker of osteogenesis, demonstrate that addition of 10 mM lithium causes earlier and higher levels of ALP expression in mineralizing cultures of hMSCs. Data are presented in from two seperate donors. (**b**) Extraction and colorimetric quantification of the calcium-binding dye, Alizarin Red S (a marker of mineralization), confirms that lithium increases the rate of osteogenic differentiation by hMSCs. P values versus appropriate "supplemented medium" control: $P < 0.05$ (*), $P < 0.01$ (**) for $n = 6$. Panel c: Addition of 500 ng mL^{-1} recombinant Dkk-1 to mineralizing hMSCs in the presence of osteogeic supplements, β-glycerophosphate, ascorbate, and bone morphogenic protein 2 disrupts the Wnt-ediated differentiation process. Histograms refer to alkaline phosphatase activity per cell. (Data reproduced with permission from Gregory *et al.*[30])

MSCS, DKK-1 AND CANCER

There are surprising parallels in Dkk-1 expression and activity between MSCs and some osteosarcoma cell lines. When assayed by RT-PCR, the osteosarcoma cell lines MG-63 and SAOS both expressed high levels of Dkk-1 in culture.[10] Furthermore, the MG-63 cell line exhibited a characteristic lag phase induced by fresh medium that could be extended by addition of the blocking Dkk-1 antiserum.[10] The role of Dkk-1 in osteosarcoma proliferation is probably similar to its role in MSCs, but this remains to be tested. Nevertheless, high Dkk-1 expression seems to be a hallmark of most osteosarcomas since screening of serum from newly diagnosed juveniles by ELISA revealed consistently high systemic levels of Dkk-1 in affected individuals (Gregory *et al.* in preparation). One interesting possibility is that Dkk-1 may be inhibiting terminal differentiation of the hMSCs, which could result in the lack of mineralization and weakening of the bone seen in some cases. The significance of high Dkk-1 levels in cases of osteosarcoma remains to be elucidated, but anti-Dkk-1 strategies may provide a novel and effective means for control of the-disease.

Multiple myeloma (MM) is a fatal malignancy of antibody-secreting plasma cells (PCs) and it accounts for 10% of all hematologic malignancies.[47] MM is the only hematological malignancy consistently associated with widespread osteolytic lesions (OLs) that cause a debilitating degenerative bone disease. Patients with overt and symptomatic MM suffer from intractable bone pain at the site of the OLs, particularly in the spine and the long bones. In advanced cases, patients suffer spinal cord compression and loss of mobility. There is no satisfactory treatment for OLs, but in extreme cases, lesions are filled with a hardening resin that strengthens the surrounding bone. OLs appear adjacent to clusters of malignant PCs, suggesting that the MM cells secrete local factors that affect the osteoclasts and osteoblasts and their regulation of normal coupled bone turnover. Initial studies have shown that the OLs occur in response to hyperactivated osteoclasts resulting from the secretion of numerous ligands secreted by the malignancy.[48] However, recent studies have utilized global gene expression profiling to identify molecular determinants of OLs in newly diagnosed MM patients. Of the genes found to be reproducibly overexpressed in cases of MM exhibiting osteolytic lesions, Dkk-1 was found to be the only secreted product.[49] Further investigation strongly correlated Dkk-1 serum levels with MRI-diagnosed osteolytic lesions. In addition, *in vitro* experiments demonstrated that Dkk-1 could reduce BMP2-induced alkaline phosphatase activity in osteoblast progenitor cells,[49] an observation confirmed by our laboratory with MSCs. It was therefore hypothesized that the abnormally high level of Dkk-1 probably prevents Wnt-mediated terminal differentiation of progenitors into osteoblasts, further disrupting the balance of normal coupled bone turnover, resulting in the persistence of OLs. Clearly, Dkk-1 activity in individuals with multiple myeloma is a candidate target for the reduction of the frequency and size of OLs. This could be accomplished by at least two means: bypassing the action of Dkk-1 at the membrane by GSK3β inhibitors is an extremely attractive method since small molecule inhibitors are available. However, GSK3β is a multi-role enzyme in mammalian cells and its inhibition could lead to undesirable side effects, especially *in vivo*. A more specific approach involves the production of recombinant blocking antibodies to the Dkk-1 molecule or its receptor. Some progress has been made in this field with the production of peptide agonists of Dkk-1 that mimic the LRP6 binding site in Dkk-1[30] and these could be used as antigens for specific antibody production.

SUMMARY

It is becoming clear that adult stem cells, and particularly hMSCs, will be powerful tools for regenerative medicine and gene therapy in the near future. Elucidation of the role of Dkk-1 and Wnt signaling in bone repair by MSCs has yielded valuable insights into the mechanism of skeletal regeneration and how this can go wrong in malignant disease. Agents that modulate Wnt signaling and Dkk-1 activity could provide a new and valuable class of pharmaceutical agents for the enhancement of tissue repair in humans.

REFERENCES

1. FRIEDENSTEIN, A.J., R.K. CHAILAKHYAN & U.V. GERASIMOV. 1987. Bone marrow osteogenic stem cells: *in vitro* cultivation and transplantation in diffusion chambers. Cell Tissue Kinetics **20:** 263–272.
2. FRIEDENSTEIN, A.J., U. GORSKAJA & N.N. KALUGINA. 1976. Fibroblast precursors in normal and irradiated mouse hematopoietic organs. Exp. Hematol. **4:** 267–274.
3. PEREIRA, R.F., K. W. HALFORD, M. D. O'HARA, *et al.* 1995. Cultures of adherent cells from marrow can serve as long-lasting precursor cells for bone, cartilage and lung in irradiated mice. Proc. Natl. Acad. Sci. **92:** 4857–4861.
4. PITTENGER, M.F., A.M. MACKAY, S.C. BECK, *et al.* 1999. Multilineage potential of adult human mesenchymal stem cells. Science **284:** 143–147.
5. SEKIYA, I., J.T. VUORISTO, B.L. LARSON & D.J. PROCKOP (2002) In vitro cartilage formation by human adult stem cells from bone marrow stroma defines the sequence of cellular and molecular events during chondrogenesis. Proc. Natl. Acad. Sci. USA **99:** 4397–4402.
6. DEXTER T.M., E. SPOONCER, R. SCHOFIELD, *et al.* 1984. Haemopoietic stem cells and the problem of self-renewal. Blood Cells **10:** 315–339.
7. AUSTIN, T.W., G.P. SOLAR, F.C. ZIEGLER, *et al.* 1997. A role for the Wnt gene family in hematopoiesis: expansion of multilineage progenitor cells. Blood **89:** 3624–3635.
8. VAN DEN BERG, D.J., A.K. SHARMA, E. BRUNO & R. HOFFMAN. 1998. Role of members of the Wnt gene family in human hematopoiesis. Blood **92:** 3189–3202.
9. WILLERT, K., J.D. BROWN, E. DANENBERG, *et al.* 2003. Wnt proteins are lipid-modified and can act as stem cell growth factors. Nature **423:** 448–452.
10. GREGORY, C.A., H. SINGH, A.S. PERRY & D.J. PROCKOP. 2003. Wnt signaling inhibitor Dkk-1 is required for re-entry into the cell cycle of human adult stem cells from bone marrow stroma (hMSCs). J. Biol. Chem. **278:** 28067–28078.
11. REYES, M. & C.M. VERFAILLIE. 2001. Characterization of multipotent adult progenitor cells, a subpopulation of mesenchymal stem cells. Ann. N.Y. Acad. Sci. **938:** 231–233.
12. JIANG, Y., B. VAESSEN, T. LENVIK, *et al.* 2002. Multipotent progenitor cells can be isolated from postnatal murine bone marrow, muscle, and brain. Exp. Hematol. **30:** 896–904.
13. JIANG, Y., B.N. JAHAGIRDAR, R.L. REINHARDT, *et al.* 2002. Pluripotency of mesenchymal stem cells derived from adult marrow. Nature **418:** 41–49.
14. JIANG, Y., D. HENDERSON, M. BLACKSTAD, *et al.* 2003. Neuroectodermal differentiation from mouse multipotent adult progenitor cells. Proc. Natl. Acad. Sci. USA **100** (Suppl. 1): 11854–118560.
15. SPEES, J.L., S.D. OLSON, J. YLOSTALO, *et al.* 2003. Differentiation, cell fusion, and nuclear fusion during ex vivo repair of epithelium by human adult stem cells from bone marrow stroma. Proc. Natl. Acad. Sci. USA **100:** 2397–2402.
16. PROCKOP, D.J. 1997. Marrow stromal cells as stem cells for non-hematopoietic tissues. Science **276:** 711–774.
17. AZIZI, S.A., D.G. STOKES, B.J. AUGELLi, *et al.* 1998. Engraftment and migration of human bone marrow stromal cells implanted in the brains of albino rats—similarities to astrocyte grafts. Proc. Natl. Acad. Sci. USA **95:** 3908–3913.
18. HORWITZ, E.M., D.J. PROCKOP, L.A. FITZPATRICK, *et al.* 1999. Transplantability and therapeutic effects of bone marrow-derived mesenchymal cells in children with osteogenesis imperfecta. Nature Med. **5:** 309–313.
19. HORWITZ, E.M., D.J. PROCKOP, P.L. GORDON, *et al.* 2001. Clinical responses to bone marrow transplantation in children with severe osteogenesis imperfecta. Blood **97:** 1227–1231.
20. HORWITZ, E.M., P.L. GORDON, W.K. KOO, *et al.* 2002. Isolated allogeneic bone marrow-derived mesenchymal cells engraft and stimulate growth in children with osteogenesis imperfecta: implications for cell therapy of bone. Proc. Natl. Acad. Sci. **99:** 8932–8937.
21. KRAUSE, D.S., N.D. THIESE, M.I. COLLECTOR, *et al.* 2001. Multi-organ, multi-lineage engraftment by a single marrow-derived stem cell. Cell **105:** 369–377.

22. HOFSTETTER, C.P., E. J. SCHWARZ, D. HESS, *et al.* 2002. Marrow stromal cells form guiding strands in the injured spinal cord and promote recovery. Proc. Natl. Acad. Sci. USA **99**: 2199–2204.

23. KOC, O.N., J. DAY, M. NIEDER, *et al.* 2002. Allogeneic mesenchymal stem cell infusion for treatment of metachromatic leukodystrophy (MLD) and Hurler syndrome (MPS-IH). Bone Marrow Transplant **30**: 215–222.

24. PITTENGER M.F. & B.J. MARTIN. 2004. Mesenchymal stem cells and their potential as cardiac therapeutics. Circ. Res. **95**: 9–20.

25. SHAKE J.G., P.J. GRUBER, W.A. BAUMGARTNER, *et al.* 2002. Mesenchymal stem cell implantation in a swine myocardial infarct model: engraftment and functional effects. Ann. Thorac. Surg. **73**: 1919–1925.

26. OUREDNIK, J., V. OUREDNIK, W.P. LYNCH, *et al.* 2002. Neural stem cells display an inherent mechanism for rescuing dysfunctional neurons. Nat. Biotechnol. **20**: 1103–1110.

27. ZHAO, L., W.M. DUAN, M. REYES, *et al.* 2002. Human bone marrow stem cells exhibit neural phenotypes and ameliorate neurological deficits after grafting into the ischemic brain of rats. Exp. Neurol. **174**: 11–20.

28. MAHMOOD, A, D. LU & M. CHOPP. 2004. Intravenous administration of marrow stromal cells (MSCs) increases the expression of growth factors in rat brain after traumatic brain injury. J. Neurotrauma **21**: 33–39.

29. CHEN J., Y. LI, M. KATAKOWSKI, X. CHEN, *et al.* 2003. Intravenous bone marrow stromal cell therapy reduces apoptosis and promotes endogenous cell proliferation after stroke in female rat. J. Neurosci. Res. **73**: 778–786.

30. GREGORY, C.A., A.S. PERRY, E. REYES, *et al.* 2004. Dkk-1 derived synthetic peptides and lithium chloride for the control and recovery of adult stem cells from bone marrow. J. Biol. Chem. **280**: 2309–2323.

31. ANGOULVANT, D., A. CLERC, S. BENCHALAL, *et al.* 2004. Human mesenchymal stem cells suppress induction of cytotoxic response to alloantigens. Biorheology **41**: 469–476.

32. LE BLANC, K. 2003. Immunomodulatory effects of fetal and adult mesenchymal stem cells. Cytotherapy **5**: 485–489.

33. LE BLANC, K., C. TAMMIK, K. ROSENDAHL, *et al.* 2003. HLA expression and immunologic properties of differentiated and undifferentiated mesenchymal stem cells. Exp. Hematol. **31**: 890–896.

34. DEVINE, S.M., C. COBBS, M. JENNINGS, *et al.* 2003. Mesenchymal stem cells distribute to a wide range of tissues following systemic infusion into nonhuman primates. Blood **101**: 2999–3001.

35. SPEES, J.L., C.A. GREGORY, H. SINGH, *et al.* 2004. Internalized antigens must be removed to prepare hypo-immunogenic mesenchymal stem cells for cell and gene therapy. Mol. Ther. **9**: 747–756.

36. STUDENY, M., F.C. MARINI, R.E. CHAMPLIN, *et al.* 2002. Bone marrow-derived mesenchymal stem cells as vehicles for interferon-beta delivery into tumors. Cancer Res. **62**: 3603–3608.

37. BRUDER, S.P., N. JAISWAL & S.E. HAYNESWORTH. 1997. Growth kinetics, self-renewal, and the osteogenic potential of purified human mesenchymal stem cells during extensive subcultivation and following cryopreservation. J. Cell Biochem. **64**: 278–294.

38. DIGIROLAMO, C.M., D. STOKES, D. COLTER, *et al.* 1999. Propagation and senescence of human marrow stromal cells in culture: a simple colony-forming assay identifies samples with the greatest potential to propagate and differentiate. Br. J. Haematol. **107**: 275–281.

39. COLTER, D.C., R. CLASS, C.M. DIGIROLAMO & D.J. PROCKOP. 2000. Rapid expansion of recycling stem cells in cultures of plastic-adherent cells from human bone marrow. Proc. Natl. Acad. Sci. USA **96**: 7294–7299.

40. GLINKA, A.W. WU, H. DELIUS, P.A. MONAGHAN, *et al.* 1998. Dickkopf-1 is a member of a new family of secreted proteins and functions in head induction Nature **391**: 357–362.

41. FEDI, P., A. BAFICO, A. NIETO SORIA, *et al.* 1999. Isolation and biochemical characterization of the human Dkk-1 homologue, a novel inhibitor of mammalian Wnt signaling. J. Biol. Chem. **274**: 19465–19472.

42. HARTMANN, C. & C.J. TABIN. 2000. Dual roles for Wnt signaling during chondrogenesis in the chicken limb. Development **127:** 3141–3159.
43. KIKUCHI, A. 2000. Regulation of beta-catenin signaling in the Wnt pathway. Biochem. Biophys. Res. Commun. **268:** 243–248.
44. HUELSKEN, J. & W. BIRCHMEIER. 2001. New aspects of Wnt signaling pathways in higher vertebrates. Curr. Opin. Genet. Dev. **11:** 547–553.
45. BAIN, G., T. MULLER, X. WANG & J. PAPKOFF. 2003. Activated beta-catenin induces osteoblast differentiation of C3H10T1/2 cells and participates in BMP2 mediated signal transduction. Biochem. Biophys. Res. Commun. **301:** 84–91.
46. RAWADI, G., B. VAYSSIERE, F. DUNN, *et al.* 2003. BMP-2 controls alkaline phosphatase expression and osteoblast mineralization by a Wnt autocrine loop. J. Bone Miner. Res. **18:** 1842–1853.
47. SHAUGHNESSY J.D. & B. BARLOGIE. 2003. Interpreting the molecular biology and clinical behaviour of multiple myeloma in the context of global gene expression profiling. Immunol. Rev. **194:** 140–163.
48. ROODMAN, G.D. 2001. Biology of osteoclast activation in cancer. J. Clin. Oncol. **19:** 3562–3571.
49. TIAN, E., F. ZHAN, R. WALKER, *et al.* 2003. The role of Wnt signaling antagonist Dkk-1 in the development of osteolytic lesions in multiple myeloma. N. Engl. J. Med. **349:** 2483–2494.

Human Retinal Progenitor Cells Grown as Neurospheres Demonstrate Time-Dependent Changes in Neuronal and Glial Cell Fate Potential

DAVID M. GAMM,[a,d] AARON D. NELSON,[b,d] AND CLIVE N. SVENDSEN[c,d]

Department of [a]Ophthalmology and Visual Sciences, [b]Neuroscience Program, Departments of [c]Anatomy and Neurology, and the [d]Waisman Center Stem Cell Research Program, University of Wisconsin, Madison, Wisconsin 53705, USA

ABSTRACT: The spatiotemporal birth order of the seven major classes of retinal cells is highly conserved among vertebrates. During retinal development, long projection neurons (ganglion cells) are produced first from resident progenitors, followed by the appearance of retinal interneurons, photoreceptors, and Müller glia. This sequence is maintained through the complex orchestration of cell-intrinsic and cell-extrinsic events and factors, including local influences between neighboring cells. Here we asked whether cultures of human prenatal retinal cells might also yield different ratios of cell types based on gestational age and time spent *in vitro*, thus recapitulating *in vivo* development. An established chopping technique was used to passage human prenatal retinal cells as neurospheres, avoiding the use of proteases and preserving cell–cell contacts and native microenvironments present *in vivo*. Retinal neurospheres cultured in this manner demonstrated specific patterns of growth over a limited time period, possibly reflecting trends in normal retinal development. Upon differentiation, immunocytochemical analysis revealed that retinal neurospheres produce predominantly glial cells with increasing gestational age and time in culture. Conversely, the percentage of βIII tubulin-positive neurons declined over time. This provides information for optimizing culture systems aimed at the study of human retinal development and the generation of specific retinal cell types for therapeutic use or drug testing.

KEYWORDS: progenitor cell; retina, human; culture; neuron; glia; Müller cell; ganglion cell

INTRODUCTION

The ability to maintain and expand cultures of primitive or undifferentiated human retinal cells would be of value in the study of human retinogenesis and possibly the treatment of retinal degenerative disorders. There are many potential or existing

Address for correspondence: David M. Gamm, M.D., Ph.D., T607 Waisman Center, University of Wisconsin, 1500 Highland Avenue, Madison, WI 53705-2280. Voice: 608-261-1516; fax: 608-263-5267.

dgamm@wisc.edu

Ann. N.Y. Acad. Sci. 1049: 107–117 (2005). © 2005 New York Academy of Sciences.
doi: 10.1196/annals.1334.011

sources of human retinal precursor cells, including embryonic stem cells, pre- and postnatal retinal progenitor cells,[1–3] and quiescent stem cells within the adult pigmented ciliary epithelium.[4] Each of these possesses advantages and disadvantages based upon multiple factors, including ethical, practical and safety concerns, growth potential, and cell fate plasticity.

In cultures derived from prenatal retina, the capacity to yield specific retinal cell types is likely influenced by the gestational age of the donor eye and/or the length of time cells are maintained and passaged in culture. This is predicted from vertebrate retinal development, where progenitor cells progress through different stages of cell fate competency, such that certain retinal cell types are born early while others appear later.[5,6] This progression is mediated through a combination of intrinsic cellular programs and extrinsic influences, including local cell–cell interactions, which have been shown to affect cell proliferation and differentiation potential *in vitro*.[7–10] Therefore, intercellular relationships inherent to specific culture techniques could have profound effects on the capacity of prenatal retinal cells to divide and adopt particular phenotypes.

One common method to propagate stem and progenitor cells from the central nervous system is in the form of spherical cell aggregates termed neurospheres.[11] Previous attempts to culture and passage human prenatal retinal cells as neurospheres have been largely unsuccessful, with most spheres failing to reform after dissociation with proteolytic enzymes.[2] Therefore, most established culture methods require that these cells to be passaged as monolayers after initial enzyme dissociation of the retinal tissue.[1,2] Experience with the culture of human cortical progenitor cells, however, suggests that human prenatal neurosphere cultures from this brain region can be expanded for prolonged periods of time using a mechanical chopping method to establish and passage the sphere populations.[12] Such an approach obviates the need for proteolytic enzymes and theoretically preserves cell surface receptors, native cell contacts, and niches thought to be important during development.

In the present study, we tested the hypothesis that, using the aforementioned culture system optimized for cortical neurospheres, human prenatal retinal neurospheres would display improved growth characteristics and follow developmental trends similar to those observed in vertebrates *in vivo*. To this end, we first performed sphere growth assays to determine the expansion potential of this system. We then used immunocytochemistry to compare the relative expression of βIII tubulin (which labels ganglion cells, a developmentally early type of retinal neuron) and glial fibrillary acidic protein (which labels Müller glial cells, generated later in development) as a function of gestational age at initial culture and duration of time spent in culture.

METHODS

Retinal Neurosphere Culture

Human prenatal eyes (between 67 and 110 days post conception) without identifiers were obtained from the fetal tissue bank at the University of Washington and dissected in chilled sterile phosphate-buffered saline solution (PBS, pH 7.4) with 0.6% glucose. Under a dissecting microscope, attached periocular tissue was carefully removed with a jeweler's forceps and Vannas scissors. Globes were then rinsed and transected posterior to the ora serrata using the Vannas scissors and the anterior portion of the eye cup and vitreous body were discarded. The loosely adherent retina was gently removed using jeweler's forceps, taking care not to disrupt the underlying retinal pigmented epithelium. In addition, margins of anterior retinal tissue near the ciliary margin and posterior retinal tissue surrounding the optic nerve head were excluded from the dissection to avoid potential contamination from ciliary epithelium and migrating astrocytes, respectively. After several washes in PBS–glucose, retinal tissue was transferred to the lid of a 16-mm petri dish as described for passaging human prenatal cortical neurosphere cultures.[12] Sections were automatically taken through the tissue using a McIlwain tissue chopper (Stoelting, Wood Dale, IL) to generate cubic tissue sections with an approximate width of 200 μm. Sections were then washed and seeded in one or two T75 flasks containing 20 mL of growth medium, consisting of DMEM/HAMS F12 (3:1), penicillin G, streptomycin sulfate, amphotericin B (1:100; Gibco), B27 (1:50; Gibco), human recombinant FGF-2 and EGF (both at 20 ng/ml; R&D Systems), and heparin (5 μg/ml; Sigma). Flasks were placed at 37°C and 5% CO_2, and half the media was changed every two days. When the largest neurospheres reached an average diameter of 500 μm, they were passaged by chopping as described above and reseeded in an effort to maintain adequate diffusion of media constituents throughout the sphere.

Sphere Growth Assays

Neurospheres were expanded for one week (one passage) or one month (three passages), at which time multiple (minimum of 12) spheres of similar size were placed into individual wells of a 96-well round bottom plate containing growth medium. Plates were kept at 37°C and 5% CO_2, and sphere volumes were determined every 2 to 4 days for up to 18 days using a Nikon TE300 inverted microscope and Integrated Morphometry Analysis software (Metamorph™, Universal Imaging corporation, Downington, PA). Half of the media in each well was exchanged for fresh growth media every 2 days.

BrdU Incorporation

Neurospheres that had been expanded for one week or one month were placed in a single well of a 6-well plate and incubated in growth medium containing 0.2 μM bromodeoxyuridine (BrdU) for 14 hours. Neurospheres were then incubated for 15 minutes in 0.5 mL Accutase (Innovative Cell Technologies, San Diego, CA) at 37°C, washed and triturated into a single cell suspension. Live cells were quantified using trypan blue exclusion and 3×10^4 cells were plated on poly-L-lysine/laminin-coated coverslips for 2 hours at 37°C and 5% CO_2 before being fixed in ice-cold methanol for 10 minutes.

Differentiation of Retinal Neurosphere Cells

At specified times in culture (one week, one month, or two months), neurospheres were collected, dissociated, counted, and plated on poly-L-lysine/laminin-coated coverslips as previously described. Cells were allowed to differentiate in plating media (DMEM/HAMS F12 (3:1), penicillin G, streptomycin sulfate, amphotericin B (1:100), B27 (1:50)) for 7 days prior to being fixed with 4% paraformaldehyde in

PBS containing 4% sucrose for 20 min. Half of the plating media was exchanged after 3 days of differentiation. For evaluation of nestin expression in undifferentiated retinal neurospheres, dissociated cells were fixed acutely (2 hours after plating).

Immunocytochemistry

Immunocytochemical procedures were carried out as previously described.[13] In brief: Cells were blocked with 5% normal goat and/or 5% normal donkey serum and 0.2% Triton X-100 solution prior to addition of primary antibodies. For BrdU detection, fixed cells were pretreated with 2 M HCl for 20 min at 37°C and subsequently washed in 0.1 M sodium borate buffer. Primary antibodies and dilutions were as follows: βIII tubulin monoclonal IgG2b (TuJ1, 1:1000; Sigma), GFAP polyclonal (1:1000; DAKO, Carpinteria, CA), BrdU rat monoclonal IgG (1:300; Accurate Chemical, Westbury, NY), and nestin polyclonal (1:300; gift of Eugene Major). Primary antibodies were incubated for 1 hr at room temperature and washed in PBS. Secondary antibodies (1:1000) conjugated to either Alexa 488 (Molecular Probes, Eugene, OR) or Cy3 (Jackson Immunoresearch, West Grove, PA) were incubated for 45 min at room temperature. Cells were then stained with Hoechst 33258 (1:10,000; Sigma) to visualize nuclei and mounted onto glass slides using Immunotech mounting media (Marseille, France).

Cell Counts and Statistics

To count cells expressing markers of interest, digital images of immunostained cells were taken using a high-resolution SPOT RT Slider camera (Diagnostic Instruments, Sterling Heights, MI). Five to ten random fields (×40 magnification) were counted from three separate coverslips per condition. Data were analyzed using an unpaired Student's t-test and expressed as mean ± SEM.

RESULTS

Human Prenatal Retinal Neurospheres Can be Grown and Passaged in the Presence of EGF and FGF-2 by Means of a Mechanical Chopping Method

We sought to determine whether human prenatal retinal neurosphere cultures could be established and expanded without the use of proteases, which can strip cell surface receptors and disrupt native intercellular relationships. Retinas were dissected from human donor eyes of different gestational ages (67 to 110 days) and sectioned using a McIlwain tissue chopper as described for human prenatal cortex.[12] Retinal sections from the different eyes were cultured separately in flasks containing defined, serum-free media supplemented with B27 and the mitogens EGF and FGF-2. The inclusion of mitogens in the growth medium is an important component of stem and progenitor cell culture systems,[14] and EGF and FGF-2 in particular are known to be present during mammalian retinal histogenesis and can promote proliferation of progenitor cells.[15,16] The retinal sections cultured in this manner formed

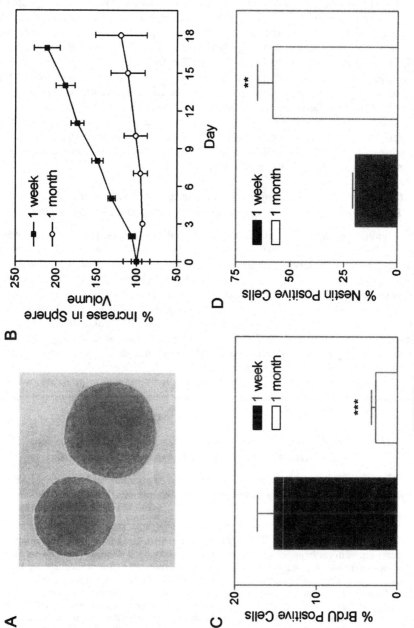

FIGURE 1. *See following page for legend.*

uniform, spherical aggregates of cells within hours after initial culture or passaging (FIG. 1A).

Individual sphere growth assays demonstrated continuous retinal neurosphere expansion at early passages (FIG. 1B). However, after three passages (equivalent to one month in culture), the majority of retinal neurospheres failed to exhibit further growth. No differences in sphere growth kinetics were observed based upon gestational age at initial culture (data not shown). The decrease in sphere growth potential over this time period correlated with a decrease in cellular incorporation of bromodeoxyuridine (BrdU) (FIG. 1C), indicating that a decline in the cell proliferation rate is at least partially responsible for the observed cessation of sphere growth. Interestingly, the expression of nestin, a marker of neural precursor cells, increased significantly over the same period (FIG. 1D). Together, these results suggest that this culture system selects for undifferentiated retinal progenitor cells that over time have the capacity to undergo a finite number of divisions.

The Neuronal and Glial Population Percentages of Cultured Human Prenatal Retinal Neurospheres Change Reciprocally with Increasing Gestational Age

Immunocytochemical analysis was employed to compare the capacity of retinal neurospheres of various gestational ages to harbor or produce two representative retinal cell types. After one month in parallel culture, retinal neurospheres derived from eyes of different gestational ages (67 to 110 days) were dissociated at the same time into single cells and plated under differentiating conditions for 7 days. Cells were then fixed and immunostained using primary antibodies directed against glial fibrillary acidic protein (GFAP), a marker for glial cells, or βIII tubulin, an early neuronal marker. As shown in FIGURE 2 (with corresponding fluorescence images available online), the percentage of cells expressing GFAP increases with advancing gestational age, whereas the percentage of βIII tubulin-positive cells decreases in a reciprocal fashion. Of note, the combined percentages do not add up to 100% owing to the existence of other retinal cell types, such as photoreceptors, that do not characteristically express either marker.

The Neuronal and Glial Cell Population Percentages of Cultured Human Prenatal Retinal Neurospheres Change over Time in Culture

Immunocytochemical analysis was also used to compare the cellular GFAP and βIII tubulin expression from differentiated human prenatal retinal neurospheres as

---

FIGURE 1. (A) Photograph (×20) of retinal neurospheres 24 hr after chopping. The spheres are uniform in appearance with a smooth surface. (B) Retinal neurosphere growth declines over time in culture. Neurosphere size is expressed as the percentage increase in sphere volume beginning at one week (*solid squares*) and one month (*open circles*) in culture. Growth assays were followed for up to 18 days. (C) Proliferation of retinal neurosphere cells declines over time in culture. Percent BrdU incorporation into neurospheres cells is shown after one week (*solid bar*) and one month (*open bar*) *in vitro*. Spheres were incubated with BrdU for 14 hr prior to dissociation and fixation (*** $P<0.0001$). (D) Nestin expression in retinal neurospheres cultures increases over time. The percentage of acutely dissociated and fixed neurosphere cells expressing nestin is shown after one week (*solid bar*) and one month (*open bar*) *in vitro* (** $P<0.01$).

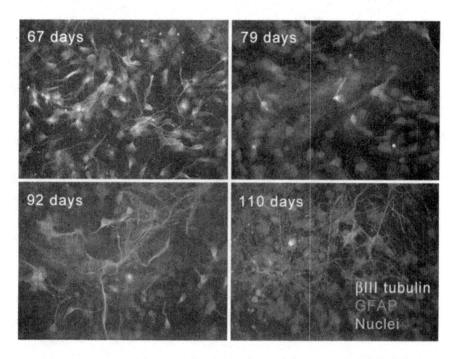

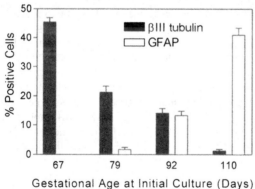

FIGURE 2. The percentage of βIII tubulin-positive (*solid bars*) and GFAP-positive (*open bars*) cell populations in differentiated retinal neurosphere cultures change reciprocally as a function of gestational age at initial culture. Retinal neurosphere cultures from four different donor eyes aged 67, 79, 92 and 110 days were followed in parallel and allowed to expand for one month. Spheres were then dissociated and incubated under differentiating conditions for 7 days prior to fixation and immunostaining. [Representative merged fluorescence images from each culture are available online.]

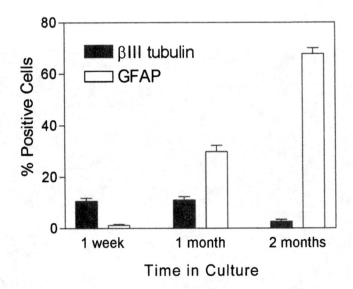

FIGURE 3. The percentage of βIII tubulin-positive (*solid bars*) and GFAP-positive (*open bars*) cell populations decrease and increase, respectively, over time *in vitro* in differentiated retinal neurosphere cultures (established from pooled retinal tissue of gestational age 87 and 91 days). At one week, one month, and two months in culture, neurospheres were dissociated, plated on coverslips and incubated under differentiating conditions for 7 days prior to fixation and imunostaining.

they aged in culture. After one week, one month, and two months spent in culture, retinal neurospheres initially established by pooling retinas from two eyes (87 and 91 days gestational age) were dissociated into single cells and plated for 7 days under differentiating conditions. This gestational age was chosen for its relatively equal representation of GFAP-positive and βIII tubulin-positive cells after 1 month *in vitro* (see FIG. 2), although some culture-to-culture variation is notable. Similar to the trend observed with retinal neurospheres of increasing gestational age at initial culture, the percentages of cells expressing GFAP or βIII tubulin under these conditions also increased or decreased, respectively, over time *in vitro* (FIG. 3). These results reveal a cumulative, time-dependent bias toward the production of glia in this culture system.

DISCUSSION

The culture method selected to propagate newborn and undifferentiated cells can have profound effects on both growth potential and the ratio of cell phenotypes produced *in vitro* and *in vivo*.[1,2,3,17] Given the importance of cell–cell interactions during development, it is perhaps not surprising that attempts to passage prenatal human retinal neurospheres by enzymatic dissociation into single cell suspensions have been largely unsuccessful.[2] A similar dilemma with human prenatal cortical neurospheres was overcome in part by passaging spheres using a mechanical chopping technique to maintain native cell contacts and minimize cell trauma.[12] Using this approach, we were able to passage human prenatal retinal neurospheres on multiple occasions. Over this time, the majority of cells within retinal neurospheres became nestin-positive, suggesting the presence of selective forces favoring the survival or expansion of undifferentiated progenitor cells and/or the elimination of more mature cell phenotypes *in vitro*. Ultimately, however, sphere growth stopped by approximately 4 weeks. Lack of sphere growth beyond one month may be related to preserved, native inhibitory influences that could also be at work in this culture system. This in turn might explain the superior growth potential observed in monolayer cultures of prenatal human retinal progenitor cells[1,2] and adult human retinal stem cells,[4] where *in vivo* cell relationships and microenvironments within the primary tissue are fully disrupted. Alternatively (or in addition), extrinsic factors may be lacking in the culture medium that are necessary to optimize retinal cell expansion. Some degree of growth limitation, however, could also result from cell-intrinsic programming that appears to be regionally specified within the developing central nervous system.[18]

In addition to the chosen culture method, two important variables to consider when characterizing cells cultured from primary tissue are the age at which the tissue is initially processed and the overall time the cells are kept in culture. This is especially true with prenatal stem and progenitor cells, which often undergo multiple changes in their proliferative capacity and cell fate potential over a relatively short time period *in vivo*.[19] Thus, strategies to generate specific cell types for therapeutic use or *in vitro* study would likely benefit from an understanding of the time course and stages of culture maturation.

In the present study, we observed a distinct trend favoring the production of GFAP-positive cells in differentiated retinal neurosphere cultures over the gestational and culture time spans examined. Concurrently, the percentage of βIII tubulin-positive cells declined under identical conditions. In our cultures, the GFAP-positive cells likely represent progenitor cell-derived Müller glia, whereas the βIII tubulin-positive cells correspond to ganglion cells. Retinal astrocytes, which also express GFAP, migrate from the optic nerve head later in development and are unlikely to contribute to the observed GFAP-positive cell populations.[20] By comparing the marker expression of an early (ganglion cell) and later (Müller glia) cell type to appear during retinogenesis, we were able to uncover clear differences in cell population percentages at multiple ages in culture. These trends are reminiscent of the changes in cell fate potential of progenitor cells during mammalian retinal development *in vivo*,[5,6] but do not reflect the observed proportions of cell phenotypes present in maturing retina. Our results showing the predominance of glial cells after neurosphere differentiation is likely the result of selective effects on cell survival, proliferation, and cell fate potential imposed by the culture system. Indeed, a limitation of this study is that it does not identify which of these effects are responsible for the cell population distribution ultimately observed.

The dynamic nature of prenatal progenitor cell cultures has also been noted in other species, tissues, and culture systems. Rat prenatal cortical progenitor cells[21] and postnatal retinal progenitor cells[22] were both noted to undergo time-dependent transitions toward a predominantly gliogenic cell fate potential. The time to first expression of photoreceptor markers was also shown to be a function of gestational age in monolayer cultures of human retinal progenitor cells.[1]

In summary, human prenatal retinal neurospheres could be passaged and expanded for up to one month using a novel culture system originally developed to optimize human prenatal cortical neurosphere growth. Furthermore, the populations of early-born retinal neurons and late-born glia emanating from these cultures decreased and increased, respectively, with advancing gestational age and time spent in culture. These findings reveal the dynamic nature of human prenatal retinal neurosphere cultures and underscore the need to consider culture age when evaluating sources of retinal cells for *in vitro* study or transplantation.

ACKNOWLEDGMENTS

This work was supported by National Institutes of Health Grant K08 EY015138 and the William Heckrodt Fund. The authors are indebted to Bernard Schneider, Ph.D., Lynda Wright, and Beth Capowski, Ph.D., for helpful discussions and technical assistance with the preparation of this manuscript.

REFERENCES

1. KELLEY, M.W., J.K. TURNER & T.A. REH. 1995. Regulation of proliferation and photo-receptor differentiation in fetal human retinal cell cultures. Invest. Ophthalmol. Vis. Sci. **36:** 1280–1289.
2. YANG, P. *et al.* 2002. In vitro isolation and expansion of human retinal progenitor cells. Exp. Neurol. **177:** 326–331.
3. KLASSEN, H. *et al.* 2004. Isolation of retinal progenitor cells from post-mortem human tissue and comparison with autologous brain progenitors. J. Neurosci. Res. **77:** 334–343.
4. COLES, B.L. *et al.* 2004. Facile isolation and the characterization of human retinal stem cells. Proc. Natl. Acad. Sci. USA **101:** 15772–15777.
5. CEPKO, C.L., C.P. AUSTIN, X. YANG, *et al.* 1996. Cell fate determination in the vertebrate retina. Proc. Natl. Acad. Sci. USA **93:** 589–595.
6. RAPAPORT, D.H., P. RAKIC & M.M. LAVAIL. 1996. Spatiotemporal gradients of cell genesis in the primate retina. Perspect. Dev. Neurobiol. **3:** 147–159.
7. BELLIVEAU, M.J. & C.L. CEPKO. 1999. Extrinsic and intrinsic factors control the genesis of amacrine and cone cells in the rat retina. Development **126:** 555–566.
8. DORSKY, R.I., D.H. RAPAPORT & W.A. HARRIS. 1995. *Xotch* inhibits cell differentiation in the *Xenopus* retina. Neuron **14:** 487–496.
9. DORSKY, R.I. *et al.* 1997. Regulation of neuronal diversity in the *Xenopus* retina by delta signaling. Nature **385:** 67–70.
10. AUSTIN, C.P. *et al.* 1995. Vertebrate retinal ganglion cells are selected from competent progenitors by the action of Notch. Trends. Genet. **7:** 403–408.
11. REYNOLDS, B.A. & S. WEISS. 1992. Generation of neurons and astrocytes from isolated cells of the adult mammalian central nervous system. Science **255:** 1707–1710.
12. SVENDSEN, C.N. 1998. A new method for the rapid and long term growth of human neural precursor cells. J. Neurosci. Meth. **85:** 141–152.
13. SUZUKI, M. *et al.* 2004. Mitotic and neurogenic effects of dehydroepiandrosterone (DHEA) on human neural stem cell cultures derived from the fetal cortex. Proc. Natl. Acad. Sci. USA **101:** 3202–3207.
14. CALDWELL, M.A. *et al.* 2001. Growth factors regulate the survival and fate of cells derived from human neurospheres. Nature Biotech. **19:** 475–479.
15. ANCHAN, R.M. *et al.* 1991. EGF and TGF-α stimulate neuroepithelial cell proliferation in vitro. Neuron **6:** 923–936.

16. LILLIEN, L. & C. CEPKO. 1992. Control of proliferation in the retina: temporal changes in responsiveness to FGF and TGF. Development **115:** 253–266.
17. JENSEN, A.M. & M.C. RAFF. 1997. Continuous observation of multipotential retinal progenitor cells in clonal density culture. Dev. Biol. **188:** 267–279.
18. OSTENFELD, T. *et al.* 2002. Regional specification of rodent and human neurospheres. Dev. Brain Res.**134:** 43–55.
19. TEMPLE, S. 2001. The development of neural stem cells. Nature **414:** 112–117.
20. SEILER, M. & R.B. ARAMANT. 1994. Photoreceptor and glial markers in human embryonic retina and in human embryonic retinal transplants to rat retina. Dev. Brain Res. **80:** 81–95.
21. CHANG, M.-Y. *et al.* 2004. Properties of cortical precursor cells cultured long term are similar to those of precursors at later developmental stages. Dev. Brain Res. **153:** 89–96.
22. ENGELHARDT, M. *et al.* 2004. The neurogenic competence of progenitors from the postnatal rat retina in vitro. Exp. Eye Res. **78:** 1025–1036.

Neural Progenitor Cell Transplants into the Developing and Mature Central Nervous System

D.S. SAKAGUCHI,[a] S.J. VAN HOFFELEN,[a,e] S.D. GROZDANIC,[b] Y.H. KWON,[c] R.H. KARDON,[c] AND M.J. YOUNG[d]

[a]Department of Genetics, Development and Cell Biology, and the Neuroscience Program, Iowa State University, Ames, Iowa 50011, USA

[b]Department of Veterinary Clinical Sciences—College of Veterinary Medicine, Iowa State University, Ames, Iowa 50011, USA

[c]Department of Ophthalmology and Visual Sciences, University of Iowa Hospitals and Clinics, Iowa City, Iowa 52242, USA

[d]Schepens Eye Research Institute, Department of Ophthalmology, Harvard Medical School, Boston, Massachusetts 02114, USA

ABSTRACT: When developing cell transplant strategies to repair the diseased or injured central nervous system (CNS), it is essential to consider host–graft interactions and how they may influence the outcome of the transplants. Recent studies have demonstrated that transplanted neural progenitor cells (NPCs) can differentiate and integrate morphologically into developing mammalian retinas. Is the ability to differentiate and to undergo structural integration into the CNS unique to specific progenitor cells, or is this plasticity a function of host environment, or both? To address these issues we have used the developing retina of the Brazilian opossum and have compared the structural integration of brain and retinal progenitor cells transplanted into the eyes at different developmental stages. The Brazilian opossum, *Monodelphis domestica*, is a small pouchless marsupial native to South America. This animal's lack of a pouch and fetal-like nature at birth circumvents the need for *in utero* surgical procedures, and thus provides an ideal environment in which to study the interactions between developing host tissues and transplanted NPCs. To test whether NPCs affect visual function we transplanted adult hippocampal progenitor cells (AHPCs) into normal, healthy adult rat eyes and performed noninvasive functional recordings. Monitoring of the retina and optic nerve over time by electroretinography and pupillometry revealed no severe perturbation in visual function in the transplant recipient eyes. Taken together, our findings suggest that the age of the host environment can strongly influence NPC differentiation and that transplantation of neural progenitor cells may be a useful strategy aimed at treating neurodegeneration and pathology of the CNS.

Address for correspondence: D.S Sakaguchi, Department of Genetics, Development and Cell Biology, 503 Science II, Iowa State University, Ames, IA 50011. Voice: 515-294-3112; fax: 515-294-8457.

dssakagu@iastate.edu

[e]Current address: Division of Biology, Kansas State University, Manhattan, KS 66506.

Ann. N.Y. Acad. Sci. 1049: 118–134 (2005). © 2005 New York Academy of Sciences.

doi: 10.1196/annals.1334.012

KEYWORDS: transplantation; retinal stem cell; retinal development; xenotransplantation; stem cells; progenitor cells

INTRODUCTION

In the mammalian central nervous system (CNS), the death of neurons is often a devastating consequence of neurodegenerative diseases. Treatments for disorders of the nervous system require methods that prevent further degeneration and also facilitate recovery of function. Stem cell transplantation offers a novel and exciting possibility for addressing these limitations.

Neural stem and progenitor cells have been isolated from various regions of the CNS, such as the hippocampus, subventricular zone, spinal cord ependyma, and retina.[1–10] Recent discoveries in the field of neural stem cell biology offer new hope for treatment of incurable neurodegenerative diseases and may also provide a potential alternative to the use of embryonic tissue. Neural stem cells are multipotential progenitors that can give rise to more specialized cells of the CNS (FIG. 1). While neural stem cell transplants have been proposed as a method of repairing the dam-

Neural and Retinal Stem Cells

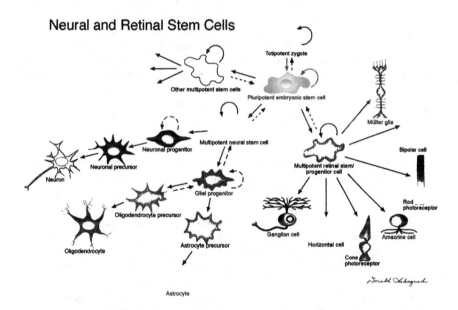

FIGURE 1. Stem cell hierarchy and multipotency of neural and retinal stem cells. Illustration depicts generation of differentiated cell types from multipotent neural and retinal stem cells. Different specialized multipotent stem cells are produced from pluripotent embryonic stem cells. Multipotent neural stem cells produce the three lineages found in the CNS. Multipotent retinal stem and progenitor cells ultimately produce the seven cell types found in the retina. *Arrows* curving back onto the same cell represent the ability for self-renewal, while the *reversed dashed arrows* represent possible reprogramming events.

aged and diseased nervous system, including the retina, it is important to note that studies examining the transplantation of "neural stem cells" may in fact be grafting mixed populations of cells, some of which may be "true" neural stem cells, but which also contain cells that are more differentiated. These cells are best-termed neural progenitor cells (NPCs) or precursor cells.[11] Neural stem/progenitor cells are defined by their ability to differentiate into cells of all CNS lineages (neurons, astrocytes, and oligodendrocytes), to self-renew, and to populate developing and/or degenerating CNS regions. Moreover, neural stem/progenitor cells possess characteristics that make them ideal vectors for brain and retinal repair. They may be expanded in culture, thereby providing a renewable supply of material for transplantation. In addition, they may be engineered to express ectopic genes for neurotrophic growth factor production that may facilitate neuroprotection.

Since the vertebrate retina has a common embryologic origin as the brain, being derived from the neural tube, it has emerged as an important model system for studying how a complex brain structure is patterned during development. Determining the molecular basis of microenvironmental cues during early retinal development will be essential for a better understanding of how the precise architectural organization of the retina is established during development. Such information can then be used in developing strategies to repair the injured retina and/or brain. During normal development the retina emerges from a pseudostratified neuroepithelium and differentiates into a highly laminated structure. Identification of different retinal cell classes in the mature retina is simplified by their segregation into separate laminae (FIG. 2). The retinal ganglion cells (RGCs), the only retinal neurons with an axon that projects out of the eye, are located in the innermost ganglion cell layer (GCL) (FIG. 2). The inner nuclear layer (INL) comprises the cell bodies of the amacrine (AC), bipolar (BP) and horizontal (HC) cells (from inner to outer retina, respectively); and the photoreceptors, the rods (R) and cones (C), are situated in the outer nuclear layer (ONL). The intrinsic glial cells in the retina, the Müller cells (MC), also have their cell bodies located in the middle region of the INL. The three nuclear layers are separated by synaptic zones referred to as the inner and outer plexiform layers (IPL and OPL, respectively) (FIG. 2).

In the mammalian retina, the death of specific cell populations is associated with a number of blinding diseases including retinitis pigmentosa, macular degeneration and glaucoma. Transplantation of NPCs into the diseased or injured eye to replace lost cellular elements or to act as support cells to facilitate recovery of disease-affected cells may become a practical strategy to treat blinding diseases in the future. A number of recent studies have begun to test the viability of intravitreal and subretinal cell transplantation in the retinas of various species using stem or progenitor cells from ocular as well as non-ocular sources.[8,10,12–24] Retinal stem and progenitor cells have been isolated from embryonic and neonatal retinas, as well as from adult ciliary epithelium.[7,9,10,25] *In vitro* studies have revealed that cells derived from retinal stem/progenitors, in some cases, appear to be capable of adopting multiple cellular fates, although the generation of photoreceptors remain somewhat more problematic. Nevertheless, these cells serve as useful candidates for retinal transplantation. Transplants of brain-derived progenitors may also be a useful strategy for retinal transplantation. Indeed, some of the most convincing examples of integration of transplanted cells into the retina were achieved using brain-derived progenitors.[12,16,18,19]

Recently we have begun to examine the influence of the host microenvironment upon the survival, differentiation, and morphologic integration of NPCs transplanted into the developing and mature retina. Using a novel experimental system we are able to investigate how the age of the host environment influences host–graft interactions and how they may influence the outcome of the transplants.

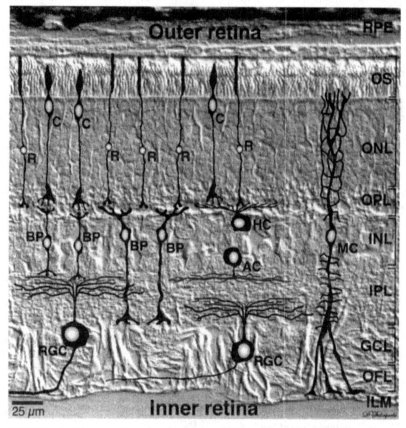

FIGURE 2. Structure of the retina. Diagram illustrates the seven basic cell types in the vertebrate retina. The retinal ganglion cells (RGCs) are situated in the inner retina with their somata in the ganglion cell layer (GCL) and their axons forming the optic fiber layer (OFL). The amacrine (AC), bipolar (BP), and horizontal (HC) cells are located in the inner nuclear layer (INL). The photoreceptors, the rods (R) and cones (C), are located in the outer nuclear layer (ONL). The cell bodies of the radially oriented Müller glial cells (MC) are located in the INL. Synaptic interactions occur within the inner and outer plexiform layers (IPL and OPL, respectively). A three-neuron path—photoreceptor to bipolar to ganglion cell—provides the direct route for the transmission of visual information. Lateral interactions, which help shape the receptive field properties of the ganglion cells, are mediated by the horizontal (HC) and amacrine (AC) cells. RPE: retinal pigment epithelium; OS: outer segments of the photoreceptors.

THE VISUAL SYSTEM OF THE BRAZILIAN OPOSSUM: AN *IN VIVO* EXPERIMENTAL MODEL SYSTEM TO STUDY STEM/PROGENITOR CELL PLASTICITY

Marsupials like the Brazilian opossum are metatherian mammals and are phylogenetically distinct from eutherian (placental) species. Marsupials and eutherians are closely related to one another, more so than to other vertebrate model species such as fish, amphibians, and birds.[26] As such, the marsupial/eutherian relationship represents a unique transitional midpoint in phylogeny relative to existing mammalian and non-mammalian vertebrate models.

Monodelphis is a small pouchless marsupial native to South America and is widely used as a model organism for comparative research on a broad range of topics relevant to human development, physiology, and disease states.[27] These animals are relatively easily maintained under laboratory conditions (they are also referred to as the laboratory opossum) and their young are born in an extremely immature, fetal-like state after a 14-day gestation (compared to 19 and 21 days' gestation for mice and rats, respectively).[28,29] Litters can vary in size from 2–13 pups and the animals reach reproductive maturity at about 6 months of age.[29]

The visual system of the Brazilian opossum possesses unique advantages for experimental analysis compared to other mammalian visual systems. Birth-dating studies have shown that the majority of cytogenesis within the retina occurs postnatally in *Monodelphis*.[28,30] Thus, the *Monodelphis* retina and visual system can serve as an *in vivo* experimental preparation, in which "embryologic" manipulations can be conducted in a developing mammalian model system without the need for *in utero* surgical procedures.[18]

IN VITRO DIFFERENTIATION OF NEURAL PROGENITOR CELLS

In vitro analysis of NPCs is essential for a more complete understanding of their differentiation potentials. We have examined murine brain progenitor cells (BPCs) and murine retinal progenitor cells (RPCs) that were isolated from neonatal brains and retinas of enhanced green fluorescent protein (GFP)–expressing transgenic mice (TgN[β-act-eGFP]04Obs)[31] as reported by Klassen and colleagues.[24] The cultures were labeled with specific antibodies to characterize their molecular differentiation. *In vitro* analysis has verified that these NPCs, when cultured under their respective differentiation conditions, were capable of expressing proteins associated with multiple cell lineages present in the CNS.[9,16,18,24]

IN VIVO ANALYSIS OF NEURAL PROGENITOR CELLS FOLLOWING TRANSPLANTATION INTO THE BRAZILIAN OPOSSUM EYE: SURVIVAL AND DIFFERENTIATION

For *in vivo* analyses, progenitor cells were transplanted by intraocular injection into the eyes of recipients of different developing and adult ages. All animal procedures for this study were in accord with the ARVO statement for the Use of Animals in Ophthalmic and Vision Research and had the approval of the Iowa State Univer-

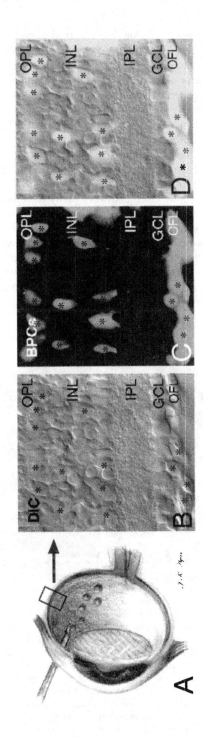

FIGURE 3. Transplantation of progenitor cells into the eye. (**A**) Neural progenitor cells were transplanted by intraocular injection into the posterior segment of the eye. (**B**) DIC image of a retinal section from a transplanted eye. (**C**) Fluorescent image of the retinal section in (**B**). *Asterisk* indicates GFP-expressing transplanted cells. (**D**) Combined DIC and fluorescent image revealing the location of the transplanted cells within the retinal architecture. OFL: optic fiber layer; GCL: ganglion cell layer; IPL: inner plexiform layer; INL: inner nuclear layer; OPL: outer plexiform layer; DIC: differential interference contrast microscopy; BPCs: brain progenitor cells.

sity Committee on Animal Care. Animals received intraocular injections of NPCs through the dorsolateral aspect of the eye by means of a beveled glass micropipette. One to two microliters of cell suspension (~50,000 cells/μL) were slowly injected into the posterior segment of each eye (FIG. 3A). An aliquot of cells used for each transplant session was plated into a sterile culture dish and visualized using fluorescence microscopy to verify the viability of the transplanted cells. By comparing "fetal-like" host environments (15 PN and younger) with the more mature cellular environments found in the older hosts (35 PN and older) we were able to investigate the influence of the host cellular microenvironment on transplanted BPCs and RPCs *in vivo*. Thus, we could compare the properties of different NPC populations within a common host microenvironment.

The xenografted NPCs survived and were easily identified after transplantation owing to their GFP expression (FIGS. 3 and 4). Our results demonstrate that these brain- and retinal-derived progenitor cells were capable of survival following xenotransplantation, even in the absence of immunosuppression. It is likely that this may in part be due to the relative purity of these CNS progenitors, which lack antigen-presenting cells and passenger leukocytes that would be present in conventional grafts of neural tissue, as well as the low level of MHC expression generally exhibited by CNS stem cells.[32–34]

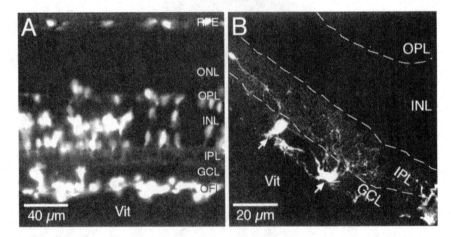

FIGURE 4. Brain and retinal progenitor cells survive, integrate and differentiate after transplantation into the developing mammalian retina. (**A**) Fluorescent image of a retinal section from an eye that received a transplant of brain progenitors at 10PN and the eye tissue examined 28 days later. Extensive integration and morphologic differentiation was observed. Transplanted cells were located along the inner retina and integrated into the GCL, INL, and ONL. GFP-expressing processes were elaborated in the IPL and OPL. (**B**) Confocal image of a retinal section from an eye that received a transplant of retinal progenitors at 10PN and the eye tissue examined 28 days later. Two transplanted cells are located in the GCL with extensive processes within the IPL. Vit: vitreal chamber; OFL: optic fiber layer; GCL: ganglion cell layer; IPL: inner plexiform layer; INL: inner nuclear layer; OPL: outer plexiform layer; ONL: outer nuclear layer; RPE: retinal pigment epithelium.

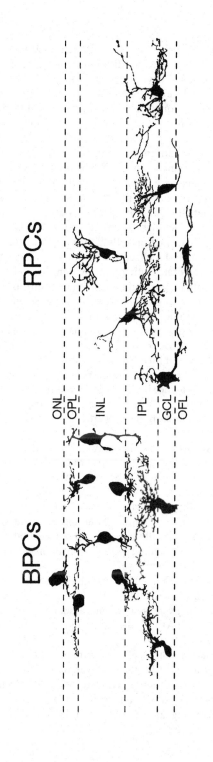

FIGURE 5. Morphologic differentiation and integration of brain progenitor cells (BPCs—*left* side of figure) and retinal progenitor cells (RPCs—*right* side of figure) 28 days following transplant. Reconstructions of GFP-expressing cells integrated into the retina after transplantation. Each cell was reconstructed from a series of confocal images with cells montaged and displayed in a single field. OFL: optic fiber layer; GCL: ganglion cell layer; IPL: inner plexiform layer; INL: inner nuclear layer; OPL: outer plexiform layer; ONL: outer nuclear layer.

Although transplanted cells survived, differentiated morphologically, and integrated into host tissue, we observed differences when the NPCs were grafted into different ages of host retinal environments. Transplanted cells were often found in the posterior segment of the eye. The NPCs were observed within the vitreous cavity, adjacent to the lens, and along the inner limiting membrane (ILM). Occasionally, transplanted cells were observed at the periphery of the eye at the iris and ciliary marginal zone.

Extensive morphologic differentiation and integration was only observed after transplantation into the youngest (5–10 PN) host retinas after 3–4 weeks post transplant. While grafted cells were capable of surviving in the older host retinas, little integration was observed and therefore we focused our studies at the earlier ages.[18,30] Transplanted BPCs in the youngest hosts often displayed differentiated morphologic patterns characteristic of specific retinal cell types, including horizontal, bipolar, amacrine, and ganglion cells.[18] In a previous study, we found that transplanted BPCs were usually localized in the nuclear layers of the retina, while the GFP processes were elaborated principally within the plexiform layers.[18] In many cells, GFP processes often appeared to segregate within specific OFF and ON sublaminae of the IPL, suggesting that these cells were capable of recognizing organizational cues present within the host retina.[18] The RPCs were also capable of developing morphologic patterns similar to amacrine and ganglion cells after grafting into the young host retinas (FIGS. 4 and 5). The presence of processes within the IPL would suggest the possibility for the establishment of synaptic connections between host neurons and transplanted cells. However, verification of synaptogenesis awaits functional studies at the single cell level and/or electron microscopic analysis. Together, these results suggest that BPCs and RPCs transplanted into a developing retina were capable of migrating into specific layers and that the transplanted cells, especially the BPCs, may be capable of recognizing specific molecular cues located within the host microenvironment that may influence their differentiation.

Recent studies suggest that grafted cells are capable of responding to cues in the developing as well as the diseased CNS.[13,18,35] A number of studies have shown that transplanted cells integrate and morphologically differentiate much better in dystrophic retinas than in healthy, normal control retinas.[16,21] It is possible that the diseased CNS in some respect recapitulates an environment similar to a developing niche, expressing factors normally only present during fetal growth. Identification of these factors and the understanding of their impact on stem cell differentiation will be important for future studies.

Specific antibodies have been used in immunohistochemical analysis to help reveal the identity of transplanted NPCs. Many transplanted cells appeared to undergo neural differentiation and expressed proteins associated with neurons (MAP2ab, calretinin) and glial cells (GFAP).[18] Morphologic differentiation was often characterized by elaboration of GFP-expressing processes into the ILM, GCL, and IPL. It is interesting to note that after transplantation there was a higher incidence of RPCs differentiating along a glial-like fate compared to those expressing the neuronal marker MAP2ab.[30] In contrast, our *in vitro* studies revealed a much greater incidence of molecular differentiation along a neuronal phenotype. These results provide evidence that host-derived signals are likely to influence the differentiation of transplanted stem/progenitor cells.

INDUCTION OF PHOTORECEPTORS FROM BRAIN AND RETINAL PROGENITORS

Van Hoffelen *et al.*[18] found that some transplanted BPCs located in the ONL labeled with antibodies to recoverin, a photoreceptor-associated protein. These cells did not have a typical photoreceptor morphology, but were incorporated among the host photoreceptors in the ONL. These results suggest that the BPCs were capable of responding to some local "retinalizing" cues within the host retina. Furthermore, our recent studies provide evidence that the immature retina secretes a soluble factor that can lead to the retinalization of brain progenitor cells into rhodopsin-expressing cells, suggesting the possibility of facilitating induction towards a photoreceptor cell fate.[36] These results complement the earlier findings of Altshuler and Cepko[37] and Watanabe and Raff,[38] demonstrating that rod photoreceptor development requires a diffusible factor produced by neonatal retina.

In our recent transplant studies we rarely observed the integration of RPCs into the ONL following grafting into the vitreal chamber.[30] However, Shatos and colleagues[30] found that the RPCs grafted into the subretinal space of mechanically injured mouse retinas were capable of integrating into the ONL, and a subset of these cells were also labeled with the retina-specific marker, recoverin.[9] It is possible that there may be position-dependent local cues that are involved in regulating the development of the RPCs into photoreceptor-like cells. The placement of RPCs into the subretinal space would likely facilitate interactions with position-dependent factors that are involved in photoreceptor differentiation. As such, it is important to consider host-derived factors and how they may influence the graft outcome in future transplant studies. The use of recipients such as the Brazilian opossum provides one with an excellent *in vivo* model system to investigate the role of these factors for neural transplantation.

NONINVASIVE FUNCTIONAL ASSAYS OF RETINA AND OPTIC NERVE FUNCTION AFTER TRANSPLANTATION OF NEURAL PROGENITOR CELLS

To test whether NPCs, which could survive and differentiate after grafting, altered visual function, we transplanted adult rat hippocampal progenitor cells (AHPCs) into normal, healthy adult rat eyes.[39] The AHPCs were clonally derived from adult Fischer 344 rats and were genetically modified to express the enhanced GFP transgene.[5] Although the multipotential AHPCs are not of retinal origin, several investigators have successfully transplanted AHPCs into rat retinas and have found that these cells can survive, differentiate, and integrate into the host tissue.[12–14,16,19,39]

The electroretinogram (ERG) and pupillary light reflex (PLR) are noninvasive methods used for determining retinal function during ophthalmic diseases and can offer objective information about the status of the retina and optic nerve.[40–43] The electroretinogram is an extracellular response produced in the retina as a result of photic stimulation. Any pathologic event, which may affect the electrophysiologic properties of the retina, can be detected by a change in ERG parameters. To register changes due to retinal pathology we have previously established standard values from healthy rat eyes.[40] The characterization of the ERG amplitudes and latency

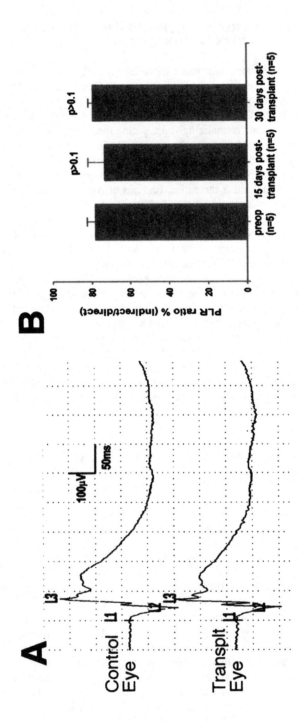

FIGURE 6. Transplantation of AHPCs into normal healthy rat eyes did not cause severe deficits in visual function. (**A**) Scotopic flash ERG recordings 30 days post transplant from one of the rats that received an AHPC transplant. *Top trace:* from the control, normal eye which did not receive AHPCs; *lower trace:* (Transplt Eye) ERG recording from the host eye that received the transplant of AHPCs. (**B**) The dynamics of the PLR response after AHPC transplantation. Monitoring of PLR amplitudes revealed no significant difference between pre- and postoperative values 15 and 30 days after AHPC transplantation.

times are likely to have significant importance in the evaluation of retinal damage and possible functional integration due to transplantation of the AHPCs. By using different protocols (flash ERG: a-wave [for outer retinal function], b-wave [for inner retinal function]; flicker ERG [for cone pathway function]), a relatively precise estimation of the location of retinal damage can be achieved. To obtain information about the status of different retinal layers after AHPC transplantation, we used scotopic flash ERG and photopic flicker ERG recordings. No significant deficits of ERG amplitudes and latency times were observed when comparing transplanted and control (non-transplanted eyes) in all tested parameters (FIG. 6A). However, the latency of onset of the a-wave was slightly, but significantly, prolonged in transplanted eyes compared to the control eyes (transplant: 23.9 ± 0.74 ms compared to control: 22.4 ± 1.81 ms, $P = 0.0479$, paired t-test, $P = 0.0479$). Since we did not detect significant integration of the neural progenitors in the ONL it is unlikely that direct contact of transplanted cells interfered with photoreceptor function, thus augmenting the a-wave latency. However, in some transplant recipient retinas we observed a radial pattern of GFAP immunoreactivity (-IR), indicative of reactive Müller glial cell labeling. This suggests the presence of a reactive response in some of the retinas after AHPC transplantation and may in part contribute to the observed slight increase in a-wave latency.

The reflex constriction of the pupil to a light stimulus, referred to as the pupillary light reflex (PLR), provides an objective measure of the afferent conduction of the visual system. The PLR involves a direct pathway to the midbrain from the retina through the optic nerve into the optic tract. The PLR has been effectively used to assess retinal and optic nerve function[40,44] and we used this technique as an additional noninvasive method for *in vivo* characterization of possible damage after AHPC transplant. Monitoring of PLR amplitudes revealed no significant difference between pre- and post-transplant values (FIG. 6B). Fifteen or 30 days after transplantation, the PLR_{ratio} was not significantly different compared to pre-transplant values ($P > 0.1$, repeated measures ANOVA with Bonferroni post-test, $n = 5$) (FIG. 6B).

PHENOTYPIC DIFFERENTIATION OF TRANSPLANTED AHPCs

Thirty days after transplantation, and after the functional recordings, the eyes were prepared for immunohistochemical analysis. The AHPCs were identified on the basis of their GFP fluorescence. GFP-expressing AHPCs were observed throughout the posterior segment of the eye. In general, retinal architecture was maintained after intravitreal transplant of AHPCs and there were no severe disturbances in retinal morphology (FIG. 7). As illustrated in FIGURE 7, retinal lamination was still evident in AHPC transplant recipient eyes (FIGS. 7A1 and B1). Transplanted AHPCs were primarily found along the inner limiting membrane (ILM), in the vitreous cavity, and around the lens, and were found as both small and large clusters and as dispersed cells (FIG. 7). In some cases transplanted cells were also found integrated into the neural retina, primarily within the inner retina (FIGS. 7A2 and B2). Cells integrated into the retina were usually present in the optic fiber layer and ganglion cell layer (GCL) and GFP-expressing processes were often observed extending from the transplanted cells in the retina (FIG. 7).

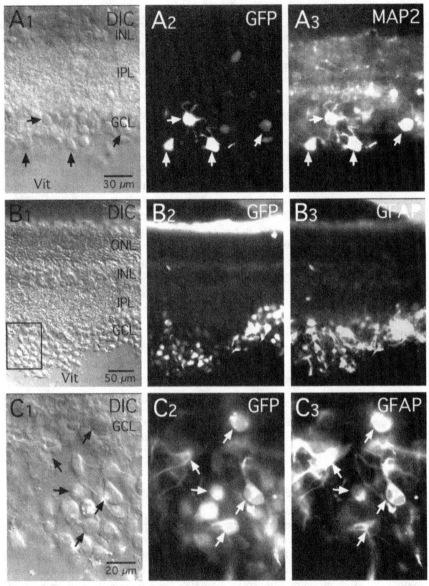

FIGURE 7. Adult hippocampal progenitor cells (AHPCs) can differentiate and integrate after transplantation into a normal, healthy eye. In each row of images, the first image is a differential interference contrast (DIC-Nomarski) image of a retinal section from an eye after AHPC transplant (A1, B1, C1); the second image in each row illustrates GFP fluorescence from the transplanted AHPCs (A2, B2, C2); and the third image in each row illustrates fluorescence from the antibody labeling visualized using Cy3 immunofluorescence (A3: MAP2, B3: GFAP, C3: GFAP). The *boxed* area in B1 is magnified in the C series of images. DIC: differential interference contrast; GFP: green fluorescent protein; MAP2: microtubule-associated protein 2; GFAP: glial fibrillary acidic protein; GCL: ganglion cell layer; IPL: inner plexiform layer; INL: inner nuclear layer; OPL: outer plexiform layer; ONL: outer nuclear layer. *Arrows* indicate examples of GFP-expressing cells co-expressing the respective antibody marker.

Antibodies directed against MAP2ab and GFAP were used to determine whether engrafted AHPCs expressed differentiated markers within the environment of the healthy eye. In the retina, MAP2ab-IR is present within the neurons of the inner retina and was observed in AHPCs integrated within the inner retina, within the ILM and GCL, and appeared to have processes extending into the inner plexiform layer (IPL) (FIG. 7A2). The presence of processes within the IPL would suggest the possibility of the establishment of synaptic connections between host neurons and transplanted cells, and awaits further analysis. An antibody against GFAP was used to determine whether grafted cells expressed this glial marker. GFAP is present in astrocytes along the inner retina and also labels Müller glial cells when they are in a reactive state. Extensive GFAP-IR was observed within GFP-expressing AHPCs located within the vitreal chamber, along and in the ILM and in the GCL (FIG. 7B1-3 and C1-3). Many of the GFP-expressing AHPCs possessed processes that were strongly immunoreactive for the GFAP antibody marker (FIG. 7).

Our study revealed a much greater incidence of transplanted AHPCs co-expressing GFAP compared to MAP2ab, suggesting that the *in vivo* environment seemed to favor the molecular differentiation of the AHPCs towards more of a glial-like fate or inhibited neuronal differentiation. Although these results do not directly address the ability of the grafted cells to integrate functionally into the retina, they show for the first time that transplanted AHPCs did not induce any dramatic perturbations in visual function observed by electroretinography and pupillometry. These results are significant for future retinal transplant studies since it is essential that grafted cells do not cause a degradation in any residual visual function.

In considering strategies for retinal repair during disease or after injury it is possible that transplanted cells may integrate into the retinal circuitry and restore visual function. In addition, it is possible that the transplanted cells may provide trophic factors that stimulate survival of remaining cells after the injury. A number of studies have demonstrated the ability of neurotrophic growth factors (BDNF, GDNF and CNTF)[45-47] to facilitate survival of retinal neurons in damaged retinas. Future studies for the treatment of retinal degenerations are likely to combine these strategies.

CONCLUSION

The retina has emerged as an important model system for investigating neural stem/progenitor cell development and plasticity. Xenotransplantation of NPCs into the developing retina may be a very useful approach to study the influence of the host microenvironment on the differentiation of transplanted cells, as well as to investigate possible reciprocal interactions between the host and grafted cells. It is clear from these results that different NPC populations can respond to the same environment in different ways. Thus, in developing NPC-based transplant models it is essential to consider not only the cell type to be used, but also the condition of the host microenvironment. The continuous monitoring of the retina and optic nerve function by electroretinography and pupillometry are powerful approaches that will be useful for evaluating visual function in future retinal transplant studies.

ACKNOWLEDGMENTS

The authors thank Dr. F. Gage (Salk Institute) for his gift of the AHPCs. Financial support was provided by the NIH (NINDS NS44007 to D.S.S.); the Hatch Act and State of Iowa Funds (D.S.S.); the Glaucoma Foundation, New York (D.S.S.); an unrestricted grant from Research to Prevent Blindness, New York, New York (Y.H.K., R.H.K.); NIH Grant 09595 (to M.J.Y.); and the Minda de Gunzburg Research Center for Retinal Transplantation (M.J.Y.). Excellent animal care was provided by the ISU Lab Animal Resources staff.

REFERENCES

1. GAGE, F.H. *et al.* 1995. Survival and differentiation of adult neuronal progenitor cells transplanted to the adult brain. Proc. Natl. Acad. Sci. USA **92:** 11879–11883.
2. WEISS, S. *et al.* 1996. Multipotent CNS stem cells are present in the adult mammalian spinal cord and ventricular neuroaxis. J. Neurosci. **16:** 7599–7609.
3. SHIHABUDDIN, L.S., T.D. PALMER & F.H. GAGE. 1999. The search for neural progenitor cells: prospects for the therapy of neurodegenerative disease. Mol. Med. Today **5:** 474–480.
4. JOHANSSON, C.B, *et al.* 1999. Identification of a neural stem cell in the adult mammalian central nervous system. Cell **96:** 25–34.
5. PALMER, T.D., J. TAKAHASHI & F.H. GAGE. 1997. The adult rat hippocampus contains primordial neural stem cells. Mol. Cell. Neurosci. **8:** 389–404.
6. ADER, M. *et al.* 2000. Formation of myelin after transplantation of neural precursor cells into the retina of young postnatal mice. Glia **30:** 301–310.
7. TROPEPE, V. *et al.* 2000. Retinal stem cells in the adult mammalian eye. Science **287:** 2032–2036.
8. CHACKO, D.M. *et al.* 2000. Survival and differentiation of cultured retinal progenitors transplanted in the subretinal space of the rat. Biochem. Biophys. Res. Commun. **268:** 842–846.
9. SHATOS, M.A. *et al.* 2001. Multipotent stem cells from the brain and retina of green mice. J. Reg. Med. **2:** 13–15.
10. COLES, B.L. *et al.* 2004. Facile isolation and the characterization of human retinal stem cells. Proc. Natl. Acad. Sci. USA **101:** 15772–15777 [e-publication October 25, 2004].
11. KLASSEN, H., D.S. SAKAGUCHI & M.J. YOUNG. 2004. Stem cells and retinal repair. Prog. Retin. Eye Res. **23:** 149–181.
12. TAKAHASHI, M. *et al.* 1998. Widespread integration and survival of adult-derived neural progenitor cells in the developing optic retina. Mol. Cell. Neurosci. **12:** 340–348.
13. NISHIDA, A. *et al.* 2000. Incorporation and differentiation of hippocampus-derived neural stem cells transplanted in injured adult rat retina. Invest. Ophthalmol. Vis. Sci. 41: 4268–4274.
14. KURIMOTO, Y. *et al.* 2001. Transplantation of adult rat hippocampus-derived neural stem cells into retina injured by transient ischemia. Neurosci. Lett. **306:** 57–60.
15. MIZUMOTO, H. *et al.* 2003. Retinal transplantation of neural progenitor cells derived from the brain of GFP transgenic mice. Vision Res. **43:** 1699–1708.
16. YOUNG, M.J. *et al.* 2000. Neuronal differentiation and morphological integration of hippocampal progenitor cells transplanted to the retina of immature and mature dystrophic rats. Mol. Cell. Neurosci. **16:** 197–205.
17. LU, B. *et al.* 2002. Transplantation of EGF-responsive neurospheres from GFP transgenic mice into the eyes of rd mice. Brain Res. **943:** 292–300.
18. VAN HOFFELEN, S.J. *et al.* 2003. Incorporation of murine brain progenitor cells into the developing mammalian retina. Invest. Ophthalmol. Vis. Sci. **44:** 426–434.
19. GUO, Y. *et al.* 2003. Engraftment of adult neural progenitor cells transplanted to rat retina injured by transient ischemia. Invest. Ophthalmol. Vis. Sci. **44:** 3194–3201.

20. PRESSMAR, S. *et al.* 2001. The fate of heterotopically grafted neural precursor cells in the normal and dystrophic adult mouse retina. Invest. Ophthalmol. Vis. Sci. **42:** 3311–3319.
21. WOJCIECHOWSKI, A.B. *et al.* 2002. Long-term survival and glial differentiation of the brain-derived precursor cell line RN33B after subretinal transplantation to adult normal rats. Stem Cells **20:** 163–173.
22. WARFVINGE, K. *et al.* 2001. Retinal integration of grafts of brain-derived precursor cell lines implanted subretinally into adult, normal rats. Exp. Neurol. **169:** 1–12.
23. CHACKO, D.M. *et al.* 2003. Transplantation of ocular stem cells: the role of injury in incorporation and differentiation of grafted cells in the retina. Vision Res. **43:** 937–946.
24. KLASSEN, H.J. *et al.* 2004. Multipotent retinal progenitors express developmental markers, differentiate into retinal neurons, and preserve light-mediated behavior. Invest. Ophthalmol. Vis. Sci. **45:** 4167–4173.
25. AHMAD, I., L. TANG & H. PHAM. 2000. Identification of neural progenitors in the adult mammalian eye. Biochem. Biophys. Res. Commun. **270:** 517–521.
26. JI, Q. *et al.* 2002. The earliest known eutherian mammal. Nature **416:** 816–822.
27. VANDEBERG, J.L. & E.S. ROBINSON. 1997. The laboratory opossum (*Monodelphis domestica*) in laboratory research. Ilar J. **38:** 4–12.
28. WEST GREENLEE, M.H. *et al.* 1996. Postnatal development and the differential expression of presynaptic terminal-associated proteins in the developing retina of the Brazilian opossum, *Monodelphis domestica*. Dev. Brain. Res. **96:** 159–172.
29. KUEHL-KOVARIK, M.C. *et al.* 1995. The gray short-tailed opossum: a novel model for mammalian development. Lab. Animal **24:** 24–29.
30. SAKAGUCHI, D.S. *et al.* 2004. Transplantation of neural progenitor cells into the developing retina of the Brazilian opossum: an in vivo system for studying stem/progenitor cell plasticity. Dev. Neurosci. **26:** 336–345.
31. OKABE, M. *et al.* 1997. "Green mice" as a source of ubiquitous green cells. FEBS Lett. **407:** 313–319.
32. KLASSEN, H. *et al.* 2003. The immunological properties of adult hippocampal progenitor cells. Vision Res. **43:** p. 947–956.
33. KLASSEN, H. *et al.* 2001. Surface markers expressed by multipotent human and mouse neural progenitor cells include tetraspanins and non-protein epitopes. Neurosci Lett. **312:** 180–182.
34. HORI, J. *et al.* 2003. Neural progenitor cells lack immunogenicity and resist destruction as allografts. Stem Cells **21:** p. 405–416.
35. WOCH, G. *et al.* 2001. Retinal transplants restore visually evoked responses in rats with photoreceptor degeneration. Invest. Ophthalmol. Vis. Sci. **42:** 1669–1676.
36. VAN HOFFELEN, S.J., M.J. YOUNG & D.S. SAKAGUCHI. 2003. Retinalization of brain-derived neural progenitor cells in vitro. Assoc. Res. Vision Ophthal. Ft. Lauderdale, FL.
37. ALTSHULER, D. & C. CEPKO. 1992. A temporally regulated, diffusible activity is required for rod photoreceptor development in vitro. Development **114:** 947–957.
38. WATANABE, T. & M.C. RAFF. 1992. Diffusible rod-promoting signals in the developing rat retina. Development **114:** 899–906.
39. SAKAGUCHI, D.S. *et al.* 2003. Transplantation of adult neural stem cells into healthy and acute ischemic rat eyes: a functional and morphological analysis. in Assoc. Res. Vision Ophthal. Ft. Lauderdale, FL.
40. GROZDANIC, S. *et al.* 2002. Characterization of the pupil light reflex, electroretinogram and tonometric parameters in healthy rat eyes. Curr. Eye Res. **25:** 69–78.
41. GROZDANIC, S.D. *et al.* 2003. Temporary elevation of the intraocular pressure by cauterization of vortex and episcleral veins in rats causes functional deficits in the retina and optic nerve. Exp. Eye Res. **77:** 27–33.
42. GROZDANIC, S.D. *et al.* 2003. Functional characterization of retina and optic nerve after acute ocular ischemia in rats. Invest. Ophthalmol. Vis. Sci. **44:** 2597–2605.
43. GROZDANIC, S.D. *et al.* 2004. Functional evaluation of retina and optic nerve in the rat model of chronic ocular hypertension. Exp. Eye Res. **79:** 75–83.

44. GROZDANIC, S. *et al.* 2003. Characterization of the pupil light reflex, electroretinogram and tonometric parameters in healthy mouse eyes. Curr. Eye Res. **26:** 371–378.
45. KO, M.L. *et al.* 2001. Patterns of retinal ganglion cell survival after brain-derived neurotrophic factor administration in hypertensive eyes of rats. Neurosci. Lett. **305:** 139–142.
46. SCHMEER, C. *et al.* 2002. Dose-dependent rescue of axotomized rat retinal ganglion cells by adenovirus-mediated expression of glial cell-line derived neurotrophic factor in vivo. Eur. J. Neurosci. **15:** 637–643.
47. WEISE, J. *et al.* 2000. Adenovirus-mediated expression of ciliary neurotrophic factor (CNTF) rescues axotomized rat retinal ganglion cells but does not support axonal regeneration in vivo. Neurobiol. Dis. **7:** 212–223.

Stem Cells for Retinal Degenerative Disorders

JASON S. MEYER,[a] MARTIN L. KATZ,[b] AND MARK D. KIRK[a]

[a]University of Missouri, Division of Biological Sciences,
Columbia, Missouri 65211, USA

[b]University of Missouri School of Medicine, Mason Eye Institute,
Columbia, Missouri 65212, USA

ABSTRACT: Many systemic and eye-specific genetic disorders are accompanied by retinal degenerations that lead to blindness. In some of these diseases retinal degeneration occurs early in life and is quite rapid, whereas in other disorders, retinal degeneration starts later and progresses very slowly. At present, no therapies are available to patients for preventing or reversing the retinal degeneration that occurs in these diseases. Implantation of neural progenitor cells into the eye may be a means by which to retard or even reverse degeneration of the retina. To evaluate the potential of neural precursor cell implantation for treating retinal degenerative disorders, neuralized mouse embryonic stem cells from green fluorescent protein (GFP) transgenic mice were administered intravitreally to normal mice, mice with early retinal degeneration, and mice with slowly progressing retinal degeneration. In normal mice, the donor cells remained in the vitreous cavity and did not associate with the host retina. In mice with early retinal degeneration, implantation of the neural precursors was performed after the degeneration was almost complete. In these animals, the donor cells primarily associated closely with the inner surface of the retina, although a small fraction of donor cells did integrate into the host retina. Donor cells implanted in mice with slowly progressing retinal degeneration also associated with the inner retinal surface, but many more of the cells integrated into the retina. These findings indicate the importance of host tissue–donor cell interactions in determining the fate of implanted neural precursor cells. These interactions will be a major consideration when devising strategies for using cell implantation therapies for neurodegenerative disorders.

KEYWORDS: embryonic stem cells; mouse; neural induction; neural progenitor; retina; transplant

INTRODUCTION

Among the symptoms of many inherited disorders is blindness due to retinal degeneration. Retinal degeneration occurs in systemic genetic disorders such as the lysosomal storage diseases neuronal ceroid-lipofuscinosis[1] and mucolipidosis type IV,[2] as well as in many diseases that lead to blindness, but have no other physiological manifestations. Among the latter disorders are retinitis pigmentosa, Leber's con-

Address for correspondence: Dr. Mark D. Kirk, University of Missouri, Division of Biological Sciences, 103 Lefevre Hall, Columbia MO 65211. Voice: 573-882-6507; fax: 573-884-5020.
  kirkm@missouri.edu

Ann. N.Y. Acad. Sci. 1049: 135–145 (2005). © 2005 New York Academy of Sciences.
doi: 10.1196/annals.1334.013

genital amaurosis and pattern dystrophy of the retina, among many others.[3–6] The
retinal degeneration that accompanies these genetic disorders can occur early in life
and progress rapidly, resulting in blindness from birth, or may have onsets as late as
adulthood and progress much more slowly.

The mutations underlying many of these genetic disorders have been identified,
which has led to substantial interest in developing gene therapy approaches for treating these diseases. Experiments with animal models have produced promising results
for the gene therapy approach for a number of retinal degenerative disorders.[7–9] Stem
cell implantation as a regenerative strategy for treating retinal disease has also become the object of extensive investigation.[10] In some cases, the gene therapy and
stem cell implantation approaches could be combined as a therapeutic approach.
Many of the investigations into stem cell implantation have employed stem cells or
neural precursor cells derived from the developed retina or brain.[10] However, embryonic stem (ES) cells may also be useful in developing cell implantation approaches
to therapy for retinal disorders. ES cells can be expanded indefinitely in culture and
are pluripotent. Under the right conditions they can be induced to differentiate into
almost any cell type, including neurons.[11–15] Thus, ES cells have the potential of
generating progeny that are optimized for their potential as therapeutic agents in
treating retinal degenerative diseases. Our studies have been directed at determining
the potential of ES cell derivatives for treating such disorders.

NEURALIZING EMBRYONIC STEM CELLS

Although ES cells have been shown to differentiate in response to environmental
cues, and such cues can be supplied by host tissues, direct implantation of undifferentiated ES cells into humans is not advisable. It has been shown that ES cells introduced into animal models have a propensity for tumor formation.[16] For therapeutic
implantation, therefore, it is important that ES cells undergo at least some degree of
differentiation prior to being introduced into the target tissue, while at the same time
retaining the capacity to respond to host tissue cues. Most approaches for cell-based
therapies for neurodegenerative disorders have employed relatively undifferentiated
cells of a neural lineage. For example, brain-derived neural precursor cells have been
used in studies to assess the potential of cell implantation for treating retinal degenerative diseases.[17–19] ES cells can be induced *in vitro* to differentiate into neural precursor or progenitor cells that can then be used for implantation in therapeutic trials.
Advantages of using ES cell–derived neural precursors over tissue-derived cells for
such trials include the ability to conduct numerous studies using an identical cell line
and the ability to precisely control the degree of differentiation of the donor cells.

A number of approaches have been developed for inducing ES cells to differentiate along a neural lineage *in vitro*, although none of these approaches results in a
homogeneous population of a single cell type.[14,20,21] Among these approaches is the
4–/4+ protocol developed by Bain and colleagues.[13] We have adopted this technique
to generate neuralized donor cells from mouse ES cells for implantation studies.[14]
Mouse ES cells are maintained in an undifferentiated state by inclusion of leukemia
inhibitory factor (LIF) in the culture medium. Induction of differentiation is initiated
by removal of LIF and plating of ES cells in non-adherent dishes to allow for the formation of embryoid bodies (EBs). After 4 days of culture in the absence of LIF, re-

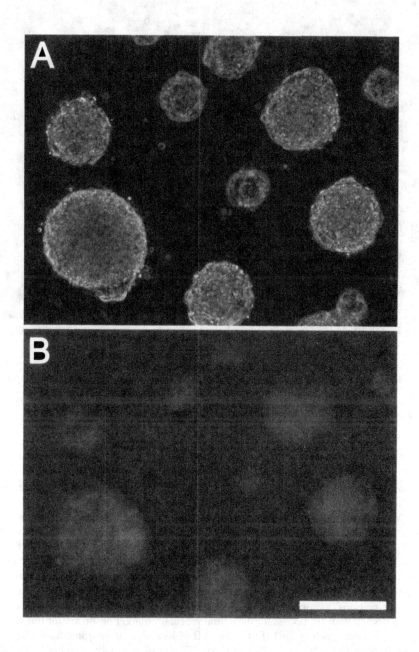

FIGURE 1. Phase-contrast (**A**) and fluorescence (**B**) micrographs of EBs composed of mouse GFP-expressing ES cells at the end of the 4–/4+ induction protocol. Scale bar in **A** and **B** equals 400 μm.

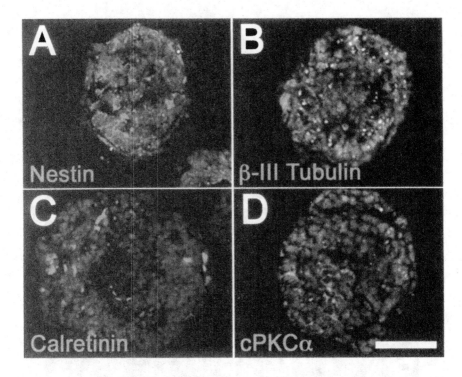

FIGURE 2. Fluorescence micrographs of EB frozen sections labeled at the end of the 4–/4+ induction protocol with anti-nestin (**A**), anti-Tuj1 (**B**), anti-calretinin (**C**), or anti-cPKCα (**D**) antibodies. Red fluorescence indicates antibody labeling. All cell nuclei in the sections were labeled with non-specific nuclear Hoechst stain (blue fluorescence). Scale bar in all panels equals 200 μm. [Illustration in color on line.]

tinoic acid is added to the culture medium. After an additional 4 days in culture, the cells are dissociated and implanted into the eyes of recipient mice.

We have extensively characterized the cells generated by this process.[14] Throughout the induction process, the donor cells are maintained in culture as unattached EBs (FIG. 1). The EBs are then dissociated to a nearly single cell suspension just prior to implantation. To characterize the state of differentiation at the end of the induction period, the EBs have been probed with antibodies to marker proteins for a variety of neural lineage cell types, including neural progenitors, immature neurons, mature neurons, astrocytes, and oligodendrocytes.[14] A great majority of the cells were labeled with antibodies for the neural precursor marker nestin and/or immature neuron markers such as Tuj1 (FIG. 2A and B). More specific neuronal markers such as calretinin (FIG. 2C) and cPKCα (FIG. 2D) are occasionally observed as well. Thus, although there was heterogeneity among the cell types generated by the 4–/4+ induction protocol, the great majority of cells used for implantation appeared to be either neural precursors or immature neurons.

INTERACTIONS BETWEEN DONOR CELLS AND HOST TISSUE

Numerous studies have been performed to characterize the interactions between neural precursor cells from various sources and the retina after donor cell implantation into the eye.[10] A focus of our work has been determining the degree to which interactions between ES cell–derived neural precursors and the retina *in vivo* is determined by the health status of the retina. We hypothesized that an actively degenerating retina would release factors that would induce migration of donor cells into the retina where they would respond to environmental cues by undergoing terminal

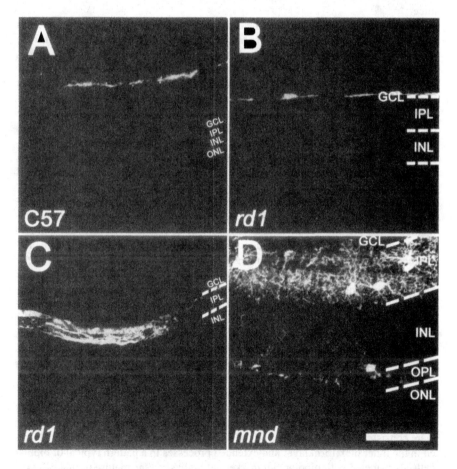

FIGURE 3. Fluorescence micrographs of GFP-expressing donor cells after implantation into the vitreous of C57BL/6J (**A**), *rd1* (**B** and **C**), and *mnd* (**D**) host mouse eyes. Scale bar equals 400 μm in **A**, 100 μm in **B** and **C**, and 50 μm in **D**. *Abbreviations:* GCL, ganglion cell layer; IPL, inner plexiform layer; INL, inner nuclear layer; OPL, outer plexiform layer; ONL, outer nuclear layer.

differentiation. These factors would not be released from normal retinas or from retinas in which the degeneration had progressed to completion. To test this hypothesis, we implanted enhanced green fluorescent protein (EGFP)–expressing mouse ES cell–derived neural precursors into the vitreous of eyes of normal C57BL/6J mice, *rd1* mutant mice with early and rapid retinal degeneration, and *mnd* mice with slowly progressive retinal degeneration.[14,22] The *rd1* mice have a mutation in the rod cGMP phosphodiesterase beta subunit gene that results in almost complete degeneration of the retinal rod photoreceptors by 4 weeks of age.[23] The *mnd* mice have a mutation in the *CLN8* gene that results in a slow continuous dropout of photoreceptor cells from the retina over the entire lifespans of these animals.[24] All implants were performed in 5-week-old mice. Implantation consisted of injection of 45,000 donor cells suspended in 1.5 μL of culture medium into the vitreous of anesthetized mice.[14,22]

In the normal C57BL/6J mice, the donor cells aggregated into sheets or clumps of cells in the vitreous (FIG. 3A). There was no evidence of any association of these cells with the host retina; they survived long-term in the vitreous (at least 6 weeks), but did not migrate into the retina. In the *rd1* mice, on the other hand, the donor cells migrated to the inner surface of the retina over a period of weeks. By 6 weeks post implantation, the great majority of donor cells were adherent to the inner retinal surface of *rd1* mice (FIG. 3B). A small fraction of the donor cells also integrated into the neural retina (FIG. 3C). These cells exhibited numerous neuron-like processes (FIG. 3C). In the *rd1* mice, the donor cells that integrated into the retina rarely penetrated deeper into the retina than the inner plexiform layer. Donor cells survived for at least 16 weeks in the recipient mouse eyes after implantation, with no signs of proliferation.

In the *mnd* mice the extent of donor cell integration into the retina was substantially higher than in the *rd1* animals (FIG. 3D).[22] While many of the donor cells remained along the inner retinal surface, a substantial fraction of them had migrated into the host retina. At least some donor cells were observed in all layers of the retina, although the majority were in the ganglion cell and inner plexiform layers. Almost all of the cells that had integrated into the retina exhibited neuronal-like morphologies, including extending numerous branching processes into the plexiform layers. The morphological appearances of the donor cells were often appropriate to the layers of the retina in which the donor cell bodies were located. For example, donor cells with nuclei in the inner nuclear layer had patterns of processes typical of amacrine and bipolar cells that are normally found in this layer (FIG. 4). Thus, many donor cells appeared to have differentiated into mature retinal neurons on the basis of their locations in the retina and the patterns of processes they extended. To further assess whether donor cells had become mature retinal neurons, sections of the retinas from treated mice were probed with antibodies against a variety of markers for specific types of mature retinal neurons. The immunolabeling data were consistent with the morphological findings. For example, cells that were located in the inner plexiform layer and extended processes in a pattern typical of bipolar cells also labeled with bipolar cell marker antibodies.[22] One mature retinal cell type that the donor cells did not appear to differentiate into were photoreceptor cells. Although a few donor cells were observed to have migrated to the outer nuclear layer of the retina, none of these cells took on the morphological appearance of mature photoreceptors, nor did they label with photoreceptor cell-specific antibodies.[22]

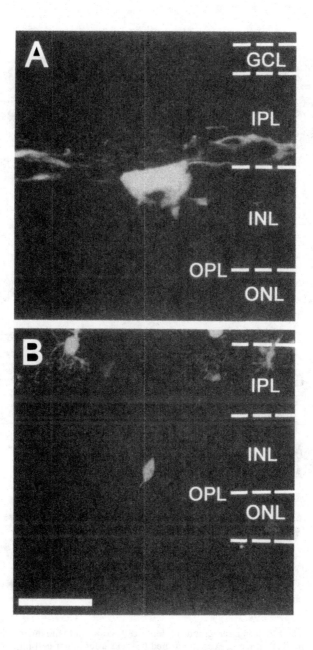

FIGURE 4. Fluorescence micrographs of donor cells that have integrated into the *mnd* host retina and exhibit morphologies similar to amacrine (**A**) and bipolar (**B**) cells. Scale bar equals 25 μm in **A** and 50 μm in **B**. *Abbreviations:* GCL, ganglion cell layer; IPL, inner plexiform layer; INL, inner nuclear layer; OPL, outer plexiform layer; ONL, outer nuclear layer.

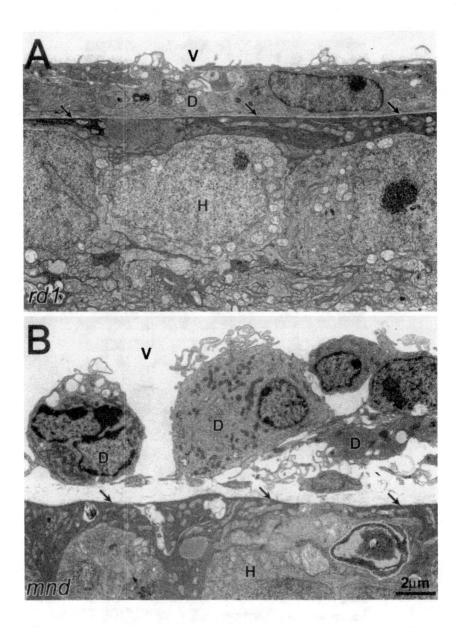

FIGURE 5. Electron micrographs of donor cells adjacent to the inner retinal surface in *rd1* (**A**) and *mnd* (**B**) mice in eyes enucleated 6 weeks after donor cell implantation. In the *rd1* mice donor cells (D) were tightly apposed to the inner surface (*arrows*) of the host retina (H). In the *mnd* mice, on the other hand, the donor cells were much more loosely associated with the inner retinal surface (*arrows*). V, vitreous cavity of the eye. Scale bar in **B** indicates magnification for both micrographs.

Not only were there differences in the degree of donor cell integration into the retinas of *mnd* and *rd1* mice, but also the nature of the association of the donor cells with the inner retinal surface was very different between these two host strains. In the *rd1* mice, the donor cells associated with the inner retinal surface were very closely adherent to the inner limiting membrane of the retina. Usually the donor cells were separated from the host retina only by the basement membrane of the Müller cells (FIG. 5A).[14] In the *mnd* mice, on the other hand, the donor cells were much more loosely associated with the inner retinal surface, and often occurred in clumps with multiple layers of donor cells (FIG. 5B).[22] In the latter strain, using electron microscopy, we also observed what appeared to be donor cells migrating into the host retina through breaks in the inner limiting membrane (data not shown). As indicated earlier, in normal C57BL/6J mice, the donor cells did not appear to associate with the host retina at all.

From these studies, it is clear that interactions between the host retina and the donor cells are highly dependent on the health status of the host retina. A normal healthy retina does not appear to send any type of signal to the donor cells that would attract these cells toward or to associate in any way with the host retina. In the eyes of *rd1* mice, in which the degenerative process was almost complete at the time of donor cell implantation, the donor cells are attracted to the retinal surface, but once there, most of them adhered strongly to this surface and ceased migration. The small number of cells that did enter into the host retina did not penetrate deeper than the inner plexiform layer. However, once in the retina, they appeared to differentiate into mature neurons with morphological features appropriate to their location in the retina. We interpret these results as follows. It appears that there are signals emanating from the *rd1* retina that induce migration of donor cells into the retina, but that these signals are relatively weak. At the same time, as the donor cells interact with the host retina, they encounter signals that induce differentiation that in turn inhibits migration. The balance between these two types of signals results in most of the donor cells remaining on the retinal surface.

In the *mnd* mice, the process of retinal degeneration is much slower than in the *rd1* animals. At the time of donor cell implantation, the retinas in the recipient *mnd* mice were actively undergoing a degenerative process. This degenerative process appears to be associated with the release of signals from the retina that induce migration of donor cells toward and into the retina. Just as in the *rd1* mice, as the donor cells interact with the host retina, they probably encounter signals that induce differentiation as well as migration. However, it appears that the relative strength of the migratory signals are greater in the *mnd* mice so that more donor cells can penetrate into the retina and penetrate more deeply before undergoing terminal differentiation that is appropriate to their location in the retina.

CONCLUSIONS

Implantation of relatively undifferentiated precursor cells into diseased tissues has great potential, in theory, to treat many degenerative disorders. Donor cell implantation has been proposed as a means of regenerating damaged host tissues or of delivering therapeutic agents. The degree to which these goals can be achieved depends a great deal on the nature of the interactions between the donor cells and the host tissues. Using intravitreal injections of the same donor cells in three strains of mice that differ in the health status of the retina, we have shown that migration of donor cells into the host tissue is highly dependent on signals emanating from the host tissue. Once they came into close contact with the host tissue, donor cells appeared to undergo differentiation and stop migrating. They appeared to differentiate into mature neurons that were appropriate to the location in the host retina to which they had migrated. These findings have important implications for the use of neural precursor cells in treating retinal degenerative disorders.

A potential limitation of this approach to therapy is that very little migration of donor cells to the photoreceptor cell layer is observed and no differentiation of donor cells into mature photoreceptor cells was observed. This is important because many retinal degenerative diseases are characterized by a primary loss of photoreceptor cells. It is possible that when donor cells traverse the retina from the vitreous side, they encounter signals to undergo terminal differentiation before they can reach the degenerating photoreceptor layer. It may be possible to overcome this limitation by implanting donor cells subretinally, directly adjacent to the photoreceptor cell layer. Such experiments remain to be undertaken.

REFERENCES

1. BENSAOULA, T. *et al.* 2000. Histopathologic and immunocytochemical analysis of the retina and ocular tissues in Batten disease. Ophthalmology **107:** 1746–1753.
2. BACH, G. 2001. Mucoliidosis type IV. Mol. Genet. Metab. **73:** 197–203.
3. MOORE, A.T. & K. EVANS. 1996. Molecular genetics of central retinal dystrophies. Aust. N.Z. J. Ophthalmol. **24:** 189–198.
4. RICHARDS, S.C. & D.J. CREEL. 1995. Pattern dystrophy and retinitis pigmentosa caused by a peripherin/RDS mutation. Retina **15:** 68–72.
5. HANEIN, S. *et al.* 2004. Leber congenital amaurosis: comprehensive survey of the genetic heterogeneity, refinement of the clinical definition, and genotype-phenotype correlations as a strategy for molecular diagnosis. Hum. Mutat. **23:** 306–317.
6. DAIGER, S.P. 2004. Identifying retinal disease gene: how far have we come, how far do we have to go? Novartis Foundation Symp. **255:** 27–36.
7. NARFSTROM, K. *et al.* 2003. In vivo gene therapy in young and adult REP65-/- dogs produces long-term visual improvement. J. Hered. **94:** 31–37.
8. SCHLICHTENBREDE, F. *et al.* 2003. Long-term evaluation of retinal function in Prph2Rd2/Rd2 mice following AAV-mediated gene replacement therapy. J. Gene Med. **5:** 757–764.
9. WELEBER, R.G., D.E. KURZ & K.M. TRZUPEK. 2003. Treatment of retinal and choroidal degenerations and dystrophies: current status and prospects for gene-based therapy. Ophthalmol. Clin. N. Amer. **16:** 583–593.
10. KLASSEN, H., D.S. SAKAGUCHI & M.J. YOUNG. 2004. Stem cells and retinal repair. Prog. Retin. Eye Res. **23:** 149–181.
11. BOHELER, K.R. *et al.* 2002. Differentiation of pluripotent embryonic stem cells into cardiomyocytes. Circ. Res. **91:** 189–220.
12. ITSKOVITZ-ELDOR, J. *et al.* 2000. Differentiation of human embryonic stem cells into embryoid bodies compromising the three embryonic germ layers. Mol. Med. **6:** 88–95.
13. BAIN, G. *et al.* 1995. Embryonic stem cells express neuronal properties in vitro. Dev. Biol. **168:** 342–357.
14. MEYER, J.S. *et al.* 2004. Neural differentiation of mouse embryonic stem cells in vitro and after transplantation into eyes of mutant mice with rapid retinal degeneration. Brain Res. **1014:** 131–144.
15. LEE, S.H. *et al.* 2000. Efficient generation of midbrain and hindbrain neurons from mouse embryonic stem cells. Nature Biotechnol. **18:** 675–679.

16. WAKITANI, S. *et al.* 2003. Embryonic stem cells injected into the mouse knee joint form teratomas and subsequently destroy the joint. Rheumatology **42:** 162–165.
17. NISHIDA, A. *et al.* 2000. Incorporation and differentiation of hippocampus-derived neural stem cells transplanted in injured adult rat retina. Invest. Ophthalmol. Vis. Sci. **41:** 4268–4274.
18. PRESSMAR, S. *et al.* 2001. The fate of heterotopically grafted neural precursor cells in the normal and dystrophic adult mouse retina. Invest. Ophthalmol. Vis. Sci. **42:** 3311–3319.
19. YOUNG, M.J., *et al.* 2000. Neuronal differentiation and morphological integration of hippocampal progenitor cells transplanted to the retina of immature and mature dystrophic rats. Mol. Cell. Neurosci. **16:** 197–205.
20. GOTTLIEB, D.J. 2002. Large-scale sources of neural stem cells. Annu. Rev. Neurosci. **25:** 381–407.
21. WERNIG, M. & O. BRUSTLE. 2002. Fifty ways to make a neuron: shifts in stem cell hierarchy and their implications for neuropathology and CNS repair. J. Neuropathol. Exp. Neurol. **61:** 101–110.
22. MEYER, J.S. *et al.* 2005. Embryonic stem cell-derived neural precursors incorporate into degenerating retina and enhance survival of host photoreceptors. Submitted for publication.
23. BOWES, C. *et al.* 1990. Retinal degeneration in the rd mouse is caused by a defect in the β subunit of rod cGMP-phosphodiesterase. Nature **347:** 677–680.
24. CHANG, B. *et al.* 1994. Retinal degeneration in motor neuron degeneration: a mouse model for ceroid-lipofuscinosis. Invest. Ophthalmol. Vis. Sci. **35:** 1071–1076.

Magnetic Resonance Tracking of Implanted Adult and Embryonic Stem Cells in Injured Brain and Spinal Cord

EVA SYKOVÁ[a,b,c] AND PAVLA JENDELOVÁ[a,b,c]

[a]Institute of Experimental Medicine ASCR, Prague, Czech Republic

[b]Center for Cell Therapy and Tissue Repair, Charles University, Second Medical Faculty, Prague, Czech Republic

[c]Department of Neuroscience, Charles University, Second Medical Faculty, Prague, Czech Republic

ABSTRACT: Stem cells are a promising tool for treating brain and spinal cord injury. Magnetic resonance imaging (MRI) provides a noninvasive method to study the fate of transplanted cells *in vivo*. We studied implanted rat bone marrow stromal cells (MSCs) and mouse embryonic stem cells (ESCs) labeled with iron-oxide nanoparticles (Endorem®) and human CD34$^+$ cells labeled with magnetic MicroBeads (Miltenyi) in rats with a cortical or spinal cord lesion. Cells were grafted intracerebrally, contralaterally to a cortical photochemical lesion, or injected intravenously. During the first week post transplantation, transplanted cells migrated to the lesion. About 3% of MSCs and ESCs differentiated into neurons, while no MSCs, but 75% of ESCs differentiated into astrocytes. Labeled MSCs, ESCs, and CD34$^+$ cells were visible in the lesion on MR images as a hypointensive signal, persisting for more than 50 days. In rats with a balloon-induced spinal cord compression lesion, intravenously injected MSCs migrated to the lesion, leading to a hypointensive MRI signal. In plantar and Basso-Beattie-Bresnehan (BBB) tests, grafted animals scored better than lesioned animals injected with saline solution. Histologic studies confirmed a decrease in lesion size. We also used 3-D polymer constructs seeded with MSCs to bridge a spinal cord lesion. Our studies demonstrate that grafted adult as well as embryonic stem cells labeled with iron-oxide nanoparticles migrate into a lesion site in brain as well as in spinal cord.

KEYWORDS: cell transplantation; magnetic resonance; contrast agents; injury; photochemical lesion; spinal cord lesion

Address for correspondence: Prof. Eva Syková M.D., D.Sc., Institute of Experimental Medicine ASCR, Vídeská 1083, 140 20 Prague 4, Czech Republic. Voice: +420-241062230; fax: + 420-241062783.

sykova@biomed.cas.cz

Ann. N.Y. Acad. Sci. 1049: 146–160 (2005). © 2005 New York Academy of Sciences.
doi: 10.1196/annals.1334.014

INTRODUCTION

The adult central nervous system (CNS) possesses a limited capacity for regeneration, and the prospects for recovery after traumatic injury or ischemia or during degenerative diseases are generally grim. Stem cell research opens new possibilities for repairing the nervous system. Transplanted stem cells can either differentiate into neural cells and replace lost populations of cells or they can produce cytokines or growth factors that can lead to neural cell rescue or enhance regeneration. Considerable progress has been made in developing effective methods for culturing different types of stem cells and transplanting them into animal models of degenerative diseases, ischemia, and brain and spinal cord injury. Although the mechanisms underlying the improvements seen in cell therapy studies remain to be determined, the results are encouraging.

The stem cells can be implanted either into the site of the lesion or into the brain ventricles or they can be administered systemically (e.g., intravenously). After transplantation, the cells integrate into the host environment and respond to intrinsic signals.[1] Stem cells administered via intravenous infusion migrate only to the lesion site and enter the nervous tissue through a more permeable or opened blood–brain barrier.[2,3]

To follow the migration and fate of implanted stem cells, the invasive analysis of brain sections post mortem has been required. Histologic methods, however, do not give us data about the dynamics of the grafting process or about the actual migration of the transplanted stem cells in the host organism. Nuclear magnetic resonance imaging (MRI) provides a noninvasive method for studying the fate of transplanted cells *in vivo* in our studies. Superparamagnetic iron-oxide nanoparticles are introduced into the cells during their cultivation prior to their grafting into the CNS tissue. The migration of transplanted adult bone marrow or embryonic stem cells can be tracked by MRI since the presence of superparamagnetic iron-oxide (SPIO) nanoparticles in the cells increases the contrast in MR images.[4–8]

SUPERPARAMAGNETIC NANOPARTICLES USED FOR CELL LABELING

Superparamagnetic contrast agents are formed by a superparamagnetic core, which is formed by iron oxide (SPIO) crystalline structures described by the general formula $Fe_2O_3 M^{2+}O$, where M is a divalent metal ion ($M = Fe^{2+}$, Mn^{2+}). For the synthesis of the contrast agents, small crystals of magnetite Fe_3O_4 are predominantly used. During the preparation of the contrast agent, the crystals are covered by a macromolecular shell, formed by dextran, starch, and other polymers, which can be chemically or biochemically modified (FIG. 1A.). On the basis of these physical properties, up to now several types of contrast agents with a nanoparticle diameter in the range of 20 to 150 nm have been described; some of them are commercially available.[9] Different modifications of the shell (e.g., the attachment of specific antibodies) enable the contrast agent to be specifically bound to the tissue.[9–12] For specific cell labeling, commercially available cell isolation kits for magnetic separation can also be used.[13–15] For magnetic separation, MACS MicroBeads, which are also superparamagnetic particles that are coupled to highly specific monoclonal antibod-

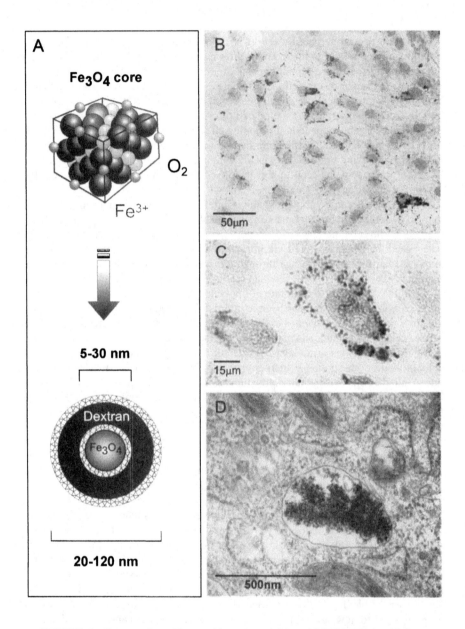

FIGURE 1. Construction of iron-oxide nanoparticles and MSC labeling with iron-oxide nanoparticles. (**A**) The contrast agent Endorem consists of a superparamagnetic Fe_3O_4 core that is coated by a dextran shell. (**B**) Cells in culture labeled with BrdU (dark nuclei), containing superparamagnetic nanoparticles (black dots). (**C**) A cell labeled with superparamagnetic nanoparticles undergoing cell division. (**D**) Transmission electron photomicrograph of a cluster of iron nanoparticles surrounded by a cell membrane. (Modified from Jendelová.[7] From: Magnetic Resonance in Medicine. ©2003 Wiley-Liss, Inc.)

ies, are used to magnetically label the target cell population. Their diameter is approximately 50 nm in size, which is comparable to the nanoparticle diameter of commonly used superparamagnetic MR contrast agents.

Nanoparticles based on magnetite are characterized by superparamagnetism, which occurs when crystal-containing regions with unpaired spins are sufficiently large that they can be regarded as thermodynamically independent, single-domain particles. The application of a nanoparticle contrast agent in MRI leads to a shortening of both T1 and T2 relaxation times that is one or two orders of magnitude greater than that seen with standard paramagnetic contrast. Thus, it is possible to observe changes in contrast on the cellular level by mini- and microimaging MR techniques.

Nanoparticles can be taken up by cells during cultivation by endocytosis,[7,8] although different methods to facilitate entry are often required (e.g., lipofection, transfection agents, antibody–iron oxide particle constructs[4,16,17]). In tissue the cells can be detected either by staining for iron to produce ferric ferrocyanide (prussian blue) or by one of a number of widely used labeling methods employed prior to transplantation (bromodeoxyuridine [BrdU], green fluorescent protein [GFP], or lacZ[2,7,18–21]).

IMPLANTATION OF MSCs IN RATS WITH A CORTICAL PHOTOCHEMICAL LESION

Marrow stromal cells (MSCs) isolated from bone marrow are multipotent progenitor cells and can differentiate in culture into osteoblasts, chondrocytes, adipocytes and myoblasts.[22] After transplantation into the brain, MSCs respond to intrinsic signals and differentiate *in vivo* into astrocytes, microglia, and even neurons.[19] The use of MSCs in cell therapies may have some advantages over the use of other sources of cells: they are relatively easy to isolate (from bone marrow), they may be used in autologous transplantation protocols, and bone marrow as a source of cells has already been approved for the treatment of hematopoietic diseases. Various routes of administration of MSCs have been tested. Kopen et al.[19] injected MSCs into the lateral ventricle, and these cells migrated throughout the forebrain and cerebellum including the striatum, the molecular layer of the hippocampus, the olfactory bulb, and the internal granular layer of the cerebellum. Li et al.[3] and Lu et al.[2] transplanted MSCs either directly into the striatum and cortex or intravenously into rats exposed to brain injury and focal cerebral ischemia. These cells migrated from the injection site and traveled to the boundary zone of the injury and the corpus callosum.

For the isolation of rat MSCs, femurs were dissected from 4-week-old Wistar rats. Marrow cells were plated on 80 cm[2] tissue culture flasks in DMEM/10% FBS with 100 U/mL penicillin and 100U/mL streptomycin. After 24 hours, the non-adherent cells were removed by replacing the medium. The medium was replaced every 2–3 days as the cells grew to confluence. After 6 to 10 passages, superparamagnetic dextran-coated iron-oxide nanoparticles, commercially known as Endorem (Guerbet, Roissy, France; 2.2 mg of iron), were added to the culture of rat MSCs 5 days prior to transplantation. After 72 hours the contrast agent was washed out and the cells were co-labeled with BrdU. Since BrdU is incorporated only into proliferating cells, double-staining for BrdU and prussian blue (staining for iron) shows that iron-oxide containing cells are viable. (FIG. 1B and C). Transmission electron microscopy confirmed the

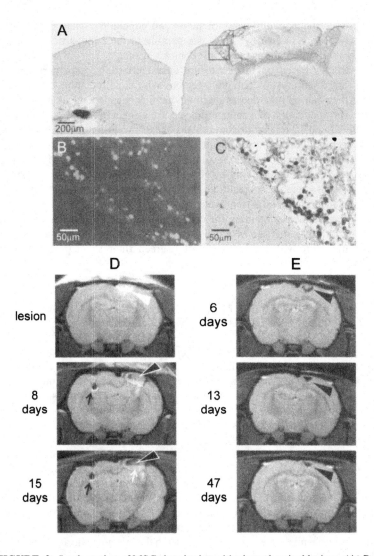

FIGURE 2. Implantation of MSCs into brains with photochemical lesions. (**A**) Prussian blue staining of an injection site in the contralateral hemisphere and a photochemical lesion 4 weeks after grafting. (**B**) Higher-magnification photomicrograph of anti-BrdU staining showing BrdU-positive MSCs in the lesion and (**C**) prussian blue-stained MSCs along the left edge of the photochemical lesion. (**D**) MR images of a cortical photochemical lesion and MSCs implanted into the contralateral hemisphere. A photochemical lesion (*white arrowhead*) 12 hours after thrombosis and prior to any cell implantation. The cell implant (*black arrow*) in the hemisphere contralateral to the lesion is seen as a hypointensive area, while the lesion (*black arrowheads*) is visible as a hypointensive signal 8 days after grafting and is further enhanced at day 15. (**E**) MR images of a cortical photochemical lesion and MSCs injected into the femoral vein. A hypointensive signal (*black arrowhead*) observed 6 days after intravenous injection was further enhanced at day 13 post injection and persisted for 47 days. (Modified from Jendelová.[7] From: Magnetic Resonance in Medicine. ©2003 Wiley-Liss, Inc.)

presence of 20–50 iron-oxide particles inside the cell, observed as membrane-bound clusters within the cell cytoplasm (FIG. 1D).

Endorem/BrdU co-labeled MSCs were grafted into rats with a cortical photochemical lesion (FIG. 2A).[7] Two weeks after implantation the cells massively populated the border zone of the damaged cortical tissue and were localized in the injured tissue around the necrotic part of the photochemical lesion (FIG. 2B and C). Only few (less than 3%) of the cells that migrated into the lesion expressed the neuronal marker NeuN when tested 28 days post implantation. No GFAP-positive cells were found in the lesion.

Rats with grafted Endorem/BrdU co-labeled cells were examined weekly for a period of 3–7 weeks post transplantation using a 4.7 T Bruker spectrometer. Single sagittal, coronal, and transversal images were obtained by a fast gradient echo sequence for localizing subsequent T_2-weighted transversal images measured by a standard turbospin echo sequence. The lesion was clearly visible on MR images 2 hours after light exposure as a hyperintense signal (FIGS. 2D and 3A) and remained visible during the entire measurement period. No recognizable hypointense signal in the lesion was detected during the first 2 days after implantation. A decrease in MR signal was found only at the injection site in animals with cells injected contralaterally to the lesion. One week after grafting, we observed a hypointense signal in the lesion, which intensified during the second and third weeks (FIG. 2D). Histologic study confirmed that a large number of prussian blue-positive cells had entered the lesion. No hypointense signal was found in other brain regions. The hypointense signal occured only in damaged areas populated with MSCs, and its intensity corresponded to prussian blue or BrdU staining.

After the intravenous injection of MSCs, we found a similar hypointense MR signal in the lesion site. The signal was observed 6 days after cell infusion and persisted for 7 weeks (FIG. 2E). Prussian blue and anti-BrdU staining confirmed the presence of iron-oxide-BrdU co-labeled cells in the lesion, which densely populated the borders of the lesion.

Only a few cells weakly stained for prussian blue were found in photochemical lesions without any implanted cells (FIG. 3D). The staining represents iron, which most likely originated in hemorrhages and iron degradation products released from iron-containing proteins (such as hemoglobin, ferritin and hemosiderin) and phagocytized by microglia/macrophages. We did not observe any BrdU-positive cells in the brains of non-grafted animals.

To mimic the signal behavior in brain tissue, we performed *in vitro* imaging of labeled cells. Rat MSCs were labeled with Endorem as described above, and a cell suspension (at concentrations of 10,000; 5000; 2500; 1250; 625 or 315 cells per µL) was suspended in 1.7% gelatin. MR images showed a hypointense signal at all concentrations above 625 cells per liter, meaning that contrast changes are visible when approximately 70 or more cells are in the image voxel.[7]

IMPLANTATION OF EMBRYONIC STEM CELLS

Embryonic stem cells (ESCs) are undifferentiated pluripotent cells derived from the inner cell mass of blastocyst-stage embryos. They maintain a normal karyotype and they can be propagated *in vitro* indefinitely in the primitive embryonic state.

ESCs can, by definition, give rise to any cell type in the body, including germ cells. Mouse ES cell-derived glial precursors, transplanted into a rat with myelin disease, interact with the host neurons to produce myelin in the brain and spinal cord.[24] Retinoic acid–treated embryoid bodies from mouse ESCs, when transplanted into a rat spinal cord 9 days after traumatic injury, differentiated into astrocytes, oligodendrocytes, and neurons and promoted recovery.[25] Bjorklund et al.[26] reported that undifferentiated mouse ESCs can become dopamine-producing neurons in the brain in a rat model of Parkinson's disease and can lead to partial functional recovery.

We used mouse D3 ESCs transfected with the pEGFP-C1 vector. Since undifferentiated ESCs may form embryonic tumors when grafted into a host animal, we induced neural differentiation by culturing eGFP ESCs in serum containing DMEM/F12 without LIF for 2 days and then transferring the cells into serum-free media supplemented with insulin, transferrin, selenium, and fibronectin for further culture. Feeder-free eGFP ESCs were labeled with Endorem (112.4mg/mL). Transplanted cells were detected by staining for iron and by GFP fluorescence. Cells were transplanted intracerebrally or intravenously on the 8th day of differentiation into adult Wistar rats with a cortical photochemical lesion.[8] When we implanted ESCs 7 days post lesion, we found a massive migration of Endorem-labeled GFP-positive cells into the lesion site regardless of the route of administration (direct injection into the contralateral hemisphere or intravenous injection; FIGS. 3E and F). In rats with a photochemical lesion and contralaterally injected cells, the cell implants were clearly visible on MR images as a hypointense area at the injection site (FIG. 3B). Two weeks after grafting, a hypointense signal was also observed in the corpus callosum. At the same time, histologic study showed that a large number of prussian blue-positive cells had entered the lesion. Many labeled cells were also detected in the corpus callosum, suggesting a migration from the contralateral hemisphere towards the lesion.

When the ESCs were injected intravenously into lesioned rats, we found a hypointense MRI signal only at the site of the lesion (FIG. 3C). The first changes in the hypointensity of the MRI signal in the lesion were observed 1 week after cell injection. The hypointense signal in the lesion reached its maximum at about 2 weeks and persisted with no apparent decrease in signal intensity for the rest of the measurement period (5 weeks). Of all the eGFP ESCs containing nanoparticles found in the lesion, 70% of them were astrocytes, very few (less than 1%) were oligodendrocytes, and about 5% of nanoparticle-labeled eGFP ESCs had differentiated into neurons.[8]

IMPLANTATION OF CD34[+] CELLS

CD34[+] cells—a subset of bone marrow cells—are known as hematopoietic progenitor cells.[27] These cells can be purified using a Miltenyi MACS sorter and clinically used for hematopoiesis restoration.[13-15] In addition, hematopoietic progenitors express neural genes,[28] and therefore they may be a potential resouce for generating neural stem cells in order to treat defects in the injured CNS. The CD34[+] cells are separated by means of immunomagnetic selection with anti-CD34 antibodies. For sorting, MicroBeads with a superparamagnetic iron-oxide core coated with a polysaccharide that is linked to an antibody are bound to the respective cell. The selected CD34[+] cells retain the magnetic label attached to their surface, and since the

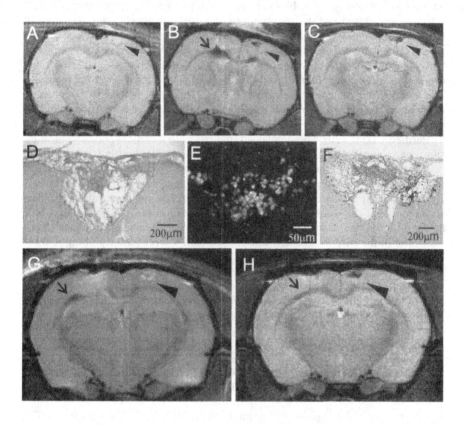

FIGURE 3. A cortical photochemical lesion and implanted eGFP ESCs or CD34+ cells. (**A**) A cortical photochemical lesion (*arrowhead*) visible on MR images two weeks after induction as a hyperintensive area with sharp hypointensive borders. (**B**) Both the cell implant of eGFP ESCs (in the hemisphere contralateral to the lesion; *arrow*) and the lesion (*arrowhead*) are hypointense in MR images 2 weeks after implantation. (**C**) A hypointensive signal in the lesion (*arrowhead*) 2 weeks after the intravenous injection of nanoparticle-labeled eGFP ESCs. (**D**) A few cells weakly stained for prussian blue were found in the photochemical lesion in animals without implanted cells. (**E**) Higher-magnification photomicrograph of GFP-labeled cells showing GFP-positive ESCs in the lesion. (**F**) Massive invasion of prussian blue-stained cells into the lesion 4 weeks after the intravenous injection of eGFP ESCs. (**G**) A cell implant of CD34+ cells (in the hemisphere contralateral to the lesion; *arrow*) visible as a hypointensive area in MR images 24 hours post injection; the lesion (*arrowhead*) remained hyperintensive. (**H**) A hypointensive signal in the lesion (*arrowhead*) 4 weeks after grafting. (**A–F** modified from Jendelová.[8] From: *Journal of Neuroscience Research* ©2004 Wiley-Liss, Inc.)

size of the MicroBeads' superparamagnetic core is comparable to the size of the superparamagnetic core of MR contrast agents (Endorem), MicroBeads provide sufficient contrast on MR images.[29]

Human cells from peripheral blood were selected by CliniMACS CD34 Selection Technology (Miltenyi). The CliniMACS is an automated cell selection device, based on MACS Technology. It enables the operator to perform large-scale magnetic cell selection in a closed and sterile system. Before selection the cells are magnetically labeled using a cell type–specific reagent. We first determined that after cryopreservation, the Microbeads remain bound to the cell surface. Transmission electron microscopy confirmed the presence of several iron-oxide nanoparticles attached to the cell surface. Iron content after mineralization was measured by spectrophotometry. The average iron content per cell was 0.275 pg. This value is lower by an order of magnitude than in the case of cell labeling using Endorem or other contrast agents that enter the cell (17 pg).[7]

Purified CD34[+] cells were implanted into rats with a cortical photochemical lesion, contralaterally to the lesion.[29] Twenty-four hours after grafting, the implanted cells were detected in the contralateral hemisphere as a hypointense area on T2W images (FIG 3G); the hypointensity of the implant decreased during the first week. At the lesion site we observed a hypointensive signal 10 days after grafting that persisted for the next 3 weeks, until the end of the experiment (FIG. 3H). Prussian blue and anti-human nuclei staining confirmed the presence of magnetically labeled human cells in the corpus callosum and in the lesion 4 weeks after grafting. CD34[+] cells were also found in the subventricular zone (SVZ). Human DNA (a human-specific 850-base-pair fragment of α-satellite DNA from human chromosome 17) was detected in brain tissue sections from the lesion using PCR, confirming the presence of human cells.[29] This is the first study showing that CD34 MicroBeads superparamagnetic nanoparticles can be used as a magnetic cell label for *in vivo* cell visualization.

IMPLANTATION OF MSCs INTO RATS WITH A SPINAL CORD COMPRESSION LESION

Severe injury of the spinal cord results in the formation of a complex lesion. The center of the lesion is necrotic, which later often results in the formation of a cyst, and the lesion is surrounded by a scar that consists of reactive astrocytes producing extracellular matrix proteins (chondroitin sulfate proteoglycans[30]). This environment prevents regenerating axons from transversing the lesion. Since it was shown by Maysinger et al.[31] that bone marrow cells secrete interleukins, stem cell factor and hematopoietic cytokine colony-stimulating factor-1 (which is a growth factor in the central nervous system) MSCs transplanted into the injured CNS could potentially stimulate tissue regeneration and promote the recovery of function and improve neurological deficits by mechanisms other than direct neuronal replacement. We therefore studied lesion size, the functional effects of cell transplantation and the fate of MSCs transplanted into rats with a spinal cord injury.

As a model of spinal cord injury we used a balloon-induced compression lesion.[32] A 2-French Fogarty catheter is inserted into the dorsal epidural space through a small hole made in the Th10 vertebral arch. A spinal cord lesion is made by balloon

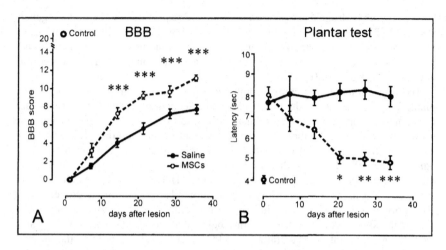

FIGURE 4. Behavioral testing of rats with spinal cord lesion. (**A**) BBB scores are significantly higher in rats with implanted MSCs than in saline-injected rats. (**B**) The sensitivity of the hind limbs is expressed by the plantar score, which is significantly higher in rats treated with MSCs.

inflation (volume 15 μL) at the Th8-Th9 spinal level. Inflation for 5 minutes produces paraplegia and is followed by gradual recovery.

MSCs labeled with Endorem were injected intravenously into the femoral vein one week after lesioning.[8] MR images were taken *ex vivo* 4 weeks after cell implantation using a standard whole-body resonator. Functional status was assessed weekly for 5 weeks after spinal cord lesioning, using the Basso–Beattie–Bresnehan (BBB) locomotor rating score and the plantar test. Our data indicate that lesioned animals with grafted MSCs have higher locomotor scores as indicated by their BBB scores and show better responses in the plantar test than do control animals (FIG. 4). Particularly, the plantar test shows the recovery of sensitivity in the hind limbs. On MR images we observed the lesion as an inhomogeneity in the tissue texture with a hypointense signal only in the area of the spinal channel (FIGS. 5A and C). Images of longitudinal as well as transversal spinal cord sections from animals grafted with magnetically labeled MSCs showed the lesion as a dark hypointense area (FIGS. 5B and D). Prussian blue staining confirmed a large number of positive cells present in the lesion. Compared to control rats, in grafted animals the lesion, which was populated by grafted MSCs, was considerably smaller, suggesting the possibility of a positive effect of the MSCs on lesion repair (FIGS. 5E and F).

HYDROGELS AND CELL–POLYMER CONSTRUCTS

Since CNS injury (particularly spinal cord injury) is accompanied by cellular death and glial scar formation that affects regeneration in the injured region, extensive research is being done to prevent scarring and to bridge the lesion. Macroporous biocompatible hydrogels can be used to eliminate scarring and to facilitate regeneration.[33,34] These hydrogels are highly biocompatible, and when implanted into nervous tissue they are known to be chemically inert and nontoxic.[35] They have a high water content (70–90%) and a very large surface area, and they are macroporous with pore sizes of 10–50 μm.

Hydrogel implantation can be combined with stem cell grafting. Before transplantation into the lesion, the hydrogels can be seeded with stem cells. In this case the hydrogels form an inert environment, allowing for the free diffusion of intrinsic growth factors, in which the cells start to differentiate and migrate. The inert environment in the hydrogels also provides an adequate standard background for MR imaging of the cells. In our studies we employed cell–polymer constructs in order to facilitate the regeneration of injured spinal cord. The scaffold of the cell–polymer constructs was made of non-resorbable biocompatible macroporous hydrogels based on copolymers of 2-hydroxyethylmethacrylate (HEMA) with communicating pores. This scaffold was seeded with Endorem-labeled MSCs (FIG. 5G). The cells migrated from the medium into the central zone of the hydrogel blocks, proving that *in vitro* the MSCs are able to move through the communicating pores of the hydrogels.[35,36]

The cells survived in the hydrogels for four weeks. The diffusion parameters of the cell–polymer constructs were determined by the TMA^+ real-time ionophoretic method using ion-selective microelectrodes.[37,38] The volume fraction represents the space between the structural components of the hydrogel, and the gel provides ingrowing cells with mechanical stability and a large surface area. The tortuosity, representing the TMA^+ molecule paths, is low in the hydrogels, since diffusion in the hydrogels is not hindered as it is in the adult CNS, particularly in injured regions.[39] The growth of MSCs in hydrogels *in vitro* did not lead to decreased diffusion in the hydrogels, and thus diffusion in the cell–polymer constructs is not hindered and allows for the free diffusion of neurotransmitters or growth factors.[36]

To evaluate the ability of cell–polymer constructs to bridge a lesion, we removed the right half of a spinal cord segment by hemisection and inserted a block of HEMA hydrogel seeded with Endorem-labeled MSCs.[40] Six weeks after implantation, the hydrogel had formed a continuous bridge between the hemisected spinal segments, re-establishing the anatomical continuity of the tissue. The hydrogel was visible on MRI as a hypointense area (FIG. 5H) and prussian blue staining confirmed positively stained cells within the hydrogel. Staining for neurofilaments (NF160 Sigma, St. Louis, MO, USA) showed axonal ingrowth into the hydrogel. Hydrogels seeded with stem cells may therefore serve as an alternative to the conventional grafting of dissociated cells, benefiting from advances in surface chemistry and the cell–cell or cell–matrix interactions that occur during development or regeneration.

FURTHER PERSPECTIVES AND CONCLUSIONS

MRI techniques can be used to monitor the fate of transplanted stem cells in a host organism. Both rat marrow stromal cells and mouse embryonic stem cells were labeled *in vitro* with superparamagnetic nanoparticles and implanted into the brain or spinal cord of rats. The distribution of the labeled cells can be monitored at regular intervals after implantation by MRI. Moreover, MicroBeads superparamagnetic

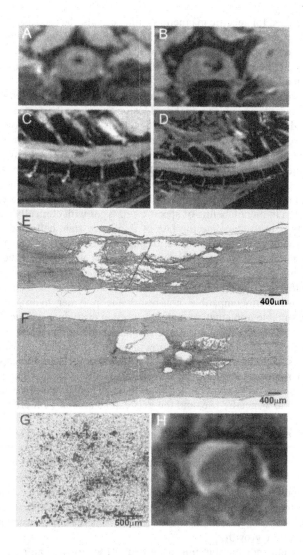

FIGURE 5. MSCs labeled with nanoparticles implanted into rats with a spinal cord compression lesion. (**A,C**) Tranversal and longitudinal sections of a spinal cord compression lesion on *ex vivo* MR images 5 weeks after compression. The lesion is seen as a hyperintensive area with a weak hypointensive signal in the spinal cord channel within the lesion. (**B,D**) Transversal and longitudinal images of a spinal cord compression lesion populated with intravenously injected nanoparticle-labeled MSCs 4 weeks after implantation. The lesion with nanoparticle-labeled cells is visible as a dark hypointensive area. (**E**) Prussian blue staining of a spinal cord compression lesion (control animal). (**F**) Prussian blue staining of a spinal cord lesion with intravenously injected nanoparticle-labeled MSCs. The lesion is populated with prussian blue-positive cells. Note the smaller lesion size in the implanted animal than in the control. (**G**) Endorem-labeled cells seeded into a hydrogel. (**H**) On MR images 6 weeks after implantation, the hydrogel was visible as a hypointensive area. (**A–F** modified from Jendelová.[8] From: *Journal of Neuroscience Research.* © 2004. Wiley-Liss, Inc.)

nanoparticles, used for specific magnetic sorting, can also be used as a magnetic cell label for *in vivo* cell visualization with MRI. Using MicroBeads, about 450 labeled cells are needed for MRI detection since the average iron content per cell is lower by an order of magnitude than in the case of intracellular contrast agents such as Endorem, where 70 cells are needed.

This MR technique gives us information about the migration speed of the transplanted cells towards a brain or spinal cord lesion and about their fate in CNS tissue. The obtained data from MR were correlated with histologic and electron microscopy findings. These correlations significantly contributed to our knowledge about the dynamics of the transplanted stem cells in the host organism. In the future, the described methodology would enable us to follow the migration of transplanted cells in humans, establish the optimal number of transplanted cells, define therapeutic windows, and monitor cell growth and possible side effects (malignancies). Currently, the described immunolabeling of specific cell types with clinically approved MicroBeads may help to elucidate the fate of implanted stem cells and evaluate the effect of cell therapy in patients with various diseases and brain or spinal cord injury.

However, in the case of large lesions, cells are not able to repair the tissue. It is necessary to fill the gap left by the lost cell population in order to provide support for tissue restoration, reduce the glial scar, and create a permissive environment for cellular ingrowth. Biocompatible non-resorbable polymer hydrogels, based on pHEMA, are suitable implantation materials used in medicine, and the implantation of these hydrogels can reduce scar tissue formation and bridge a lesion, creating an environment permissive for cellular ingrowth and the diffusion of various neuroactive substances, including growth factors.[41] These hydrogels are in many ways similar to the environment in developing nervous tissue[35] and can mechanically support the ingrowing cells and axons. In addition, their chemical and physical properties can be tailored to a specific use, and the gels can be seeded with different types of stem cells creating cell–polymer constructs.

At the present time we have to take into consideration all necessary precautions before the practical employment of cell therapy in patients suffering from degenerative diseases of the CNS or in patients with severe traumatic lesions of the CNS can be realized. While the use of a patient's autologous (adult) stem cells will probably not have serious complications, current experiments with embryonic stem cells do not give us enough information about the behavior of the transplanted cells in the host organism, especially about their influence outside the target structures and their potential neoplastic growth.

Animal studies indicate that improved neural functioning can result from bone marrow stem cell transplantation.[2,8,20,21,23] Therefore cell therapy has started to be used in patients suffering from brain and spinal cord injury. On the basis of our experimental studies, autologous bone marrow mononuclear cell implantation is being used in our Phase I clinical study in patients with transversal spinal cord lesions (Autologous Transplantation of Bone Marrow Cells into Patients with Transversal Spinal Cord Injury, coordinator E. Syková). At this time we can conclude that implantation is safe, as there were no complications following intravenous or intra-arterial administration into 9 acute and 9 chronic patients. We are using MEP, SEP, MRI, and the Asia score in our patient follow-up. Although partial improvement was observed in acute patients, further evaluation of a much larger population of patients is needed before any conclusions can be made.

REFERENCES

1. ROSARIO, C.M. *et al.* 1997. Differentiation of engrafted multipotent neural progenitors towards replacement of missing granule neurons in meander tail cerebellum may help determine the locus of mutant gene action. Development **124:** 4213–4224.
2. LU, D. *et al.* 2001. Adult bone marrow stromal cells administered intravenously to rats after traumatic brain injury migrate into brain and improve neurological outcome. Neuroreport **12:** 559–563.
3. NAKANO, K. *et al.* 2001. Differentiation of transplanted bone marrow cells in the adult mouse brain. Transplantation **71:** 1735–1740.
4. BULTE, J.W. *et al.* 1999. Neurotransplantation of magnetically labeled oligodendrocyte progenitors: magnetic resonance tracking of cell migration and myelination. Proc. Natl. Acad. Sci. USA **96:** 15256–15261.
5. BULTE, J.W. *et al.* 2001. Magnetodendrimers allow endosomal magnetic labeling and in vivo tracking of stem cells. Nat. Biotechnol. **19:** 1141–1147.
6. HOEHN, M. *et al.* 2002. Monitoring of implanted stem cell migration in vivo: A highly resolved in vivo magnetic resonance imaging investigation of experimantal stroke in rat. Proc. Natl. Acad. Sci. USA **100:** 1073–1078.
7. JENDELOVÁ, P. *et al.* 2003. Imaging the fate of implanted bone marrow stromal cells labeled with superparamagnetic nanoparticles. Magn.. Reson. Med. **50:** 767–776.
8. JENDELOVÁ, P. *et al.* 2004. MR tracking of transplanted bone marrow and embryonic stem cells labeled by iron oxide nanoparticles in rat brain and spinal cord. J. Neurosci. Res. **76:** 232–243.
9. WANG, Y.X., S.M. HUSSAIN & G.P. KRESTIN. 2001. Superparamagnetic iron oxide contrast agents: physicochemical characteristics and applications in MR imaging. Eur. Radiol. **11:** 2319–2331.
10. WEISSLEDER, R. *et al.* 1997. Magnetically labeled cells can be detected by MR imaging. J. Magn. Reson. Imaging **7:** 258–263.
11. YEH, T. *et al.* 1993. Intracellular labeling of T-cells with superparamagnetic contrast agents. Magn. Reson. Med. **30:** 617–625.
12. YEH, T. *et al.* 1995. In vivo dynamic MRI tracking of rat T-cells labeled with superparamagnetic iron-oxide particles. Magn. Reson. Med. **33:** 200–208.
13. LANG, P. *et al.* 2003. Transplantation of highly purified CD34+ progenitor cells from unrelated donors in pediatric leukemia. Blood **101:** 1630–1636.
14. HANDGRETINGER, R. *et al.* 2002. Isolation and transplantation of highly purified autologous peripheral CD34(+) progenitor cells: purging efficacy, hematopoietic reconstitution and long-term outcome in children with high-risk neuroblastoma. Bone Marrow Transplant. **29:** 731–736.
15. HANDGRETINGER, R. *et al.* 2001. Megadose transplantation of purified peripheral blood CD34(+) progenitor cells from HLA-mismatched parental donors in children. Bone Marrow Transplant. **27:** 777–783.
16. ARBAB, A.S. *et al.* 2003. Characterization of biophysical and metabolic properties of cells labeled with superparamagnetic iron oxide nanoparticles and transfection agent for cellular MR imaging. Radiology **229:** 838–846.
17. FRANK, J.A. *et al.* 2002. Magnetic intracellular labeling of mammalian cells by combining (FDA-approved) superparamagnetic iron oxide MR contrast agents and commonly used transfection agents. Acad Radiol. (Suppl. 2) **9:** S484–487.
18. BRAZELTON, T.R. *et al.* 2000. From marrow to brain: expression of neuronal phenotypes in adult mice. Science **290:** 1775–1779.
19. KOPEN, G.C., D.J. PROCKOP & D.G. PHINNEY. 1999. Marrow stromal cells migrate throughout forebrain and cerebellum, and they differentiate into astrocytes after injection into neonatal mouse brains. Proc. Natl. Acad. Sci. USA **96:** 10711–10716.
20. LU, D., *et al.* 2001. Intraarterial administration of marrow stromal cells in a rat model of traumatic brain injury. J. Neurotrauma **18:** 813–819.
21. SASAKI, M., *et al.* 2001. Transplantation of an acutely isolated bone marrow fraction repairs demyelinated adult rat spinal cord axons. Glia **35:** 26–34.
22. PROCKOP, D.J. 1997. Marrow stromal cells as stem cells for nonhematopoietic tissues. Science. **276:** 71–74.

23. LI, Y. *et al.* 2001. Treatment of stroke in rat with intracarotid administration of marrow stromal cells. Neurology. **56:** 1666–1672.
24. BRUSTLE, O. 1999. Building brains: neural chimeras in the study of nervous system development and repair. Brain Pathol. **9:** 527–545.
25. MCDONALD, J.W. *et al.* 1999. Transplanted embryonic stem cells survive, differentiate and promote recovery in injured rat spinal cord. Nat. Med. **5:** 1410–1412.
26. BJORKLUND, L.M. *et al.* 2002. Embryonic stem cells develop into functional dopaminergic neurons after transplantation in a Parkinson rat model. Proc. Natl. Acad. Sci. USA. **99:** 2344–2349.
27. CIVIN, C.I. & S.D. GORE. 1993. Antigenic analysis of hematopoiesis: a review. J. Hematother. **2:** 137–144.
28. GOOLSBY, J. *et al.* 2003. Hematopoietic progenitors express neural genes. Proc. Natl. Acad. Sci. USA **100:** 14926–14931.
29. JENDELOVÁ, P., *et al.* MR tracking of human CD34+ progenitor cells separated by means of immunomagnetic selection and transplanted into injured rat brain. Cell Transplant. In press.
30. BECKER, D., C.L. SADOWSKY & J.W. MCDONALD. 2003. Restoring function after spinal cord injury. Neurologist **9:** 1–15.
31. MAYSINGER, D., O. BEREZOVSKAYA & S. FEDOROFF. 1996. The hematopoietic cytokine colony stimulating factor 1 is also a growth factor in the CNS. II. Microencapsulated CSF-1 and LM-10 cells as delivery systems. Exp. Neurol. **141:** 47–56.
32. VANICKY, I., *et al.* 2001. A simple and reproducible model of spinal cord injury induced by epidural balloon inflation in the rat. J. Neurotrauma **18:** 1399–1407.
33. WOERLY, S. *et al.* 1998. Heterogeneous PHPMA hydrogels for tissue repair and axonal regeneration in the injured spinal cord. J. Biomater. Sci. Polym. Ed. **9:** 681–711.
34. WOERLY, S. *et al.* 2001. Reconstruction of the transected cat spinal cord following NeuroGel implantation: axonal tracing, immunohistochemical and ultrastructural studies. Int. J. Dev. Neurosci. **19:** 63–83.
35. LESNY, P. *et al.* 2002. Polymer hydrogels usable for nervous tissue repair. J. Chem. Neuroanat. **23:** 243–247.
36. LESNY, P. *et al.* 2004. Macroporous hydrogels based on 2-hydroxyethyl methacrylate. Part 4: Growth of rat bone marrow stromal cells in three-dimensional hydrogels with positive and negative surface charges and in polyelectrolyte complexes. J. Mater. Sci.: Mater. Med. In press.
37. NICHOLSON, C. & E. SYKOVA. 1998. Extracellular space structure revealed by diffusion analysis. Trends Neurosci. **21:** 207–215.
38. SYKOVÁ, E. 1997. Extracellular space volume and geometry of the rat brain after ischemia and central injury. Adv. Neurol. **73:** 121–135.
39. SYKOVÁ, E. 2004. Diffusion properties of the brain in health and disease. Neurochem. Int. **45:** 453–466.
40. JENDELOVÁ, P. *et al.* 2004. Hydrogel implantation into a spinal cord lesion: an alternative to conventional cell grafting. Program No. 106.13.2004. Abstract Viewer/Itinerary Planner <http://sfn.scholarone.com/itin2004/index.html>.
41. WOERLY, S. *et al.* 1999. Neural tissue formation within porous hydrogels implanted in brain and spinal cord lesions: ultrastructural, immunohistochemical, and diffusion studies. Tissue Eng. **5:** 467–488.

The Miniature Pig as an Animal Model in Biomedical Research

PETR VODIČKA,[a,b] KAREL SMETANA, JR.,[b,c] BARBORA DVOŘÁNKOVÁ,[b,c] TERESA EMERICK,[d] YINGZHI Z. XU,[d] JITKA OUREDNIK,[d] VÁCLAV OUREDNIK,[d] AND JAN MOTLÍK[a,b]

[a]Institute of Animal Physiology and Genetics, Department of Physiology of Reproduction, Academy of Sciences of the Czech Republic, Liběchov, the Czech Republic

[b]Center for Cell Therapy and Tissue Repair, Prague, the Czech Republic

[c]Charles University, 1st Faculty of Medicine, Institute of Anatomy, Prague, the Czech Republic

[d]Department of Biomedical Sciences, College of Veterinary Medicine, Iowa State University, Ames, Iowa 50011, USA

ABSTRACT: Crucial prerequisites for the development of safe preclinical protocols in biomedical research are suitable animal models that would allow for human-related validation of valuable research information gathered from experimentation with lower mammals. In this sense, the miniature pig, sharing many physiological similarities with humans, offers several breeding and handling advantages (when compared to non-human primates), making it an optimal species for preclinical experimentation. The present review offers several examples taken from current research in the hope of convincing the reader that the porcine animal model has gained massively in importance in biomedical research during the last few years. The adduced examples are taken from the following fields of investigation: (a) the physiology of reproduction, where pig oocytes are being used to study chromosomal abnormalities (aneuploidy) in the adult human oocyte; (b) the generation of suitable organs for xenotransplantation using transgene expression in pig tissues; (c) the skin physiology and the treatment of skin defects using cell therapy–based approaches that take advantage of similarities between pig and human epidermis; and (d) neurotransplantation using porcine neural stem cells grafted into inbred miniature pigs as an alternative model to non-human primates xenografted with human cells.

KEYWORDS: animal model; stem cell; transgenic pig; transplantation

INTRODUCTION

Animal models of human diseases have always played a central role in biomedical research for the exploration and development of new therapies. However, the

Address for correspondence: Petr Vodička, Department of Physiology of Reproduction, Institute of Animal Physiology and Genetics, Rumburská 89, Liběchov 277 21, Czech Republic. Voice: +420 315 639 580; fax: + 420 315 639 510.
vodicka@iapg.cas.cz

Ann. N.Y. Acad. Sci. 1049: 161–171 (2005). © 2005 New York Academy of Sciences.
doi: 10.1196/annals.1334.015

evolutionary gap between humans and many of the applied animal models (such as rodents) has always hampered a direct applicability of the knowledge gained for human therapy. In this regard, the pig model offers many advantages. Being a domesticated eutherian (placental) mammal, the pig has evolved similarly with humans and represents a taxon with diverse selected phenotypes.[1] On the other hand, the pig also represents an evolutionary clade distinct enough from primates and rodents to provide considerable power in the understanding of genetic complexity, that is, how genetic variation contributes to diverse phenotypes and diseases. The similarities between numerous physiological functions of pigs and humans have stimulated a wide range of biomedical research, for which, in order to simplify husbandry of pigs used in experimental work, several miniature strains have been bred. In this paper, we will consider examples from four research areas (reproduction, transgenesis, epidermal, and neural stem cell research) in which porcine models can successfully be applied.

PORCINE OOCYTES AND MODELING OF ANEUPLOIDY

One of the serious problems in human reproduction is chromosomal aneuploidy. Current estimates suggest that among 20–24-year old women, approximately 2% of all clinically recognized pregnancies involve a chromosomally abnormal fetus. This figure increases to almost one-third of all recognized pregnancies among women aged 40 years or older.[2] A substantial proportion of these chromosomal abnormalities is linked to recombination, which occurs during fetal development. Another fraction of these abnormalities can be ascribed to irregularities in chromosome processing during oocyte meiotic division.

Studies of human aneuploidy have been hampered both by the inherent difficulty of studying oocyte development and by the lack of a suitable animal model. Most previous studies have been performed using a mouse model, but in several aspects (see example below) this model does not allow a direct comparison with humans. Porcine oocytes, on the other hand, closely resemble human ones, taking into account both their morphology and the timing of meiotic maturation.[3] Mammalian oocytes are arrested at prophase I (prophase of the first meiotic division), before a surge of the preovulatory luteinizing hormone (LH) induces oocyte maturation. In mice, the resumption of meiosis starts by chromosome condensation and germinal vesicle (another term for the oocyte's nucleus) breakdown, which occurs approximately 2–5 hours after the endogenous LH peak, and metaphase II is reached 7–8 hours later (depending on mouse strain). Contrary to this, both human and porcine oocytes require about 20 hours for germinal vesicle breakdown and reach the second meiotic metaphase about 40 hours after the LH peak or *in vitro* culture. With this in mind, the plentiful source of ovaries and oocytes from prepubescent gilts (young sows) allows more extensive studies of the molecular mechanisms involved in the condensation of chromatin and subsequent chromosomal segregation. Moreover, the data obtained with oocytes isolated from prepubescent or young cycling gilts can easily be compared with the morphology and biochemistry of oocytes from aged animals. Thus, the porcine model can be used to investigate important problems associated with late maternity in humans.

TRANSGENESIS IN PIGS

Transgenic animals can serve as models for many human diseases. In mice, the technology of homologous recombination in embryonic stem cells (ESCs) provides an ideal tool for modeling genetically based disorders. Nevertheless, the relatively short lifespan of mice and the genetic and physiological differences between mice and humans makes the comparison and application of data between the two species very difficult. Transgenic pigs can serve as an ideal biomedical model bridging this gap since they are physiologically closer to humans, have a longer lifespan than mice, and are easily bred in controlled conditions.

There have been several reports describing the isolation of porcine ESCs or embryonic germ cells (EGCs), but animals created by injection of these cells into blastocysts display only somatic chimerism.[4] No transgenic cells have been detected to participate in the germ line (gamete formation), and so this methodology failed to deliver fully transgenic animals. Therefore, for a long time, the only technique for creating transgenic pigs has been pronuclear injection of a gene construct[5] (FIG. 1A). This method, however, suffers from several drawbacks, the most serious being the random integration site of the gene construct in the host genome, the lack of control over transgene expression levels, and differences in transgene expression in the offspring. These obstacles cause the costs for the production of transgenic pigs by pronuclear microinjection to be too high.[6]

Some alternative approaches addressing the above problems in transgenic pig production include sperm-mediated DNA transfer[7] (FIG. 1B) and oocyte transduction with viruses, which was first performed with retroviral vectors based on the Moloney murine leukemia virus.[8] However, retroviruses undergo epigenetic modification, and retroviral expression is shut off either during embryogenesis[8] or shortly after birth.[9] Recently, vectors derived from lentiviruses were successfully used to transduce human ESCs, as well as mouse and rat embryos.[10] The same technique was less efficient in pre-implantation monkey embryos where the transferred gene was expressed only in extraembryonic tissues.[11] When one-cell porcine embryos were injected with a recombinant lentiviral vector into the perivitelline space (FIG. 1C). Seventy percent of piglets born carried the transgene DNA, and 94% of these pigs also expressed the transgene coding for green fluorescent protein (GFP). Importantly, the transgene was transmitted through the germ line, and tissue-specific transgene expression was achieved by infection of porcine embryos with lentiviral vectors containing the human keratin K14 promoter.[12] These data indicate that lentiviral gene transfer into the porcine genome can result in a high efficiency of transgenesis and could reduce the costs of transgenic animals by a factor of ten or more, compared to pronuclear DNA microinjection.

The pig stands out as the most suitable donor for animal-to-human xenotransplantation because of the similar size and physiological capacity of its organs. Practically, however, the use of xenografted tissues of non-human origin faces many immunologic barriers. Humans and old world monkeys have natural antibodies against the $\alpha(1,3)$-galactosyl epitope, which is expressed on all porcine cells. These antibodies elicit a strong immune response in the host resulting in complement activation and hyperacute rejection of transplanted tissue. To tackle this problem, the success of pig cloning by somatic cell nucleus transfer (NT) into enucleated metaphase II oocytes has brought the possibility of using gene targeting in somatic

A Pronuclear DNA microinjection   **B** Sperm-mediated DNA transfer

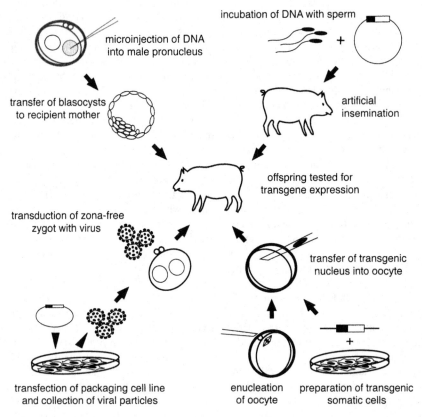

FIGURE 1. Methods available for the production of transgenic animals. (**A**) Pronuclear DNA microinjection: Recombinant DNA (transgene) is purified from a plasmid backbone and injected into the male pronucleus of a one-cell porcine embryo. Injected embryos are cultured *in vitro* up to the blastocyst stage or directly transferred into surrogate mothers. (**B**) Sperm-mediated DNA transfer: Linearized plasmid DNA is incubated with sperm (DNA binding can be enhanced by the use of antibodies against sperm surface epitopes). Sperm will internalize surface-bound DNA and can then be used for artificial insemination. (**C**) Viral oocyte transduction: The gene construct is transfected into a packaging cell line, and viral particles are collected. For successful transduction, the zona pellucida (a thick glycoprotein envelope surrounding the zygote) must be overcome. The virus is either injected into the perivitelline space between the zona and the zygote, or the zona is removed enzymatically, and the zona-free zygote is co-cultured with viral particles. The transduced embryos are transferred to surrogate mothers. (**D**) Somatic cell nuclear transfer: Somatic cells are transfected with the gene construct and selected for transgene integration (random or selection for homologous recombinants). The nuclei of transgenic cells are transferred into enucleated metaphase II oocytes. The activation of embryonic development is induced chemically or by electric current. The embryos are cultured up to the blastocyst stage *in vitro* and then transferred to surrogate mothers.

cells followed by NT for transgenic animal production (FIG. 1D). Naturally, the first gene knockout by homologous recombination in pigs involved α-1,3-galactosyl-transferase, which is responsible for the expression of the α(1,3)-galactosyl epitope on porcine cells.[13] This success finally made the genetic alterations possible that were needed to overcome the immunologic barrier between porcine and human tissue. Unfortunately, animals created by somatic cell NT display frequent abnormalities, and further improvement of nuclear reprogramming before NT is needed.[14] Thus, only efficient homologous recombination together with an optimal NT technology can secure, in the future, the low production costs that are essential for a widespread use of transgenic swine as disease models and donor animals for xenotransplantation.

EPIDERMAL STEM CELLS

Epidermal stem cells (EpSCs) are responsible for the continuous turnover of the epidermis and for its regeneration after injury. They are very slowly proliferating elements with, at least theoretically, an unlimited number of mitoses. Because of this property, EpSCs can be identified by their tendency to retain a considerable amount of labeled precursors of DNA synthesis, such as bromo-deoxyuridine or tritiated thymidine, and are thus referred to as label-retaining cells.[15] EpSCs divide preferentially by asymmetric mitosis, in which one daughter cell retains the properties of an EpSC and the other becomes a transit amplifying cell, which proliferates quickly and has a restricted number of mitoses.[16] EpSCs are located in the basal layer of the interfollicular epidermis and in the bulge region of the outer root sheath of the hair follicle.

No specific phenotypic markers of EpSCs have yet been discovered. However, expression of some molecules is detected, at least preferentially, in cells with the properties of EpSCs. These cells are strongly positive for α6β4 and β1 integrin receptors, which mediate adhesion to extracellular matrix molecules, such as collagen or laminin. Expression of integrins is downregulated during terminal differentiation of keratinocytes. On the other hand, expression of cadherins, mediators of calcium-dependent cell-to-cell adhesion, is minimal in EpSCs, but common in other epidermal cell types.[17] From the large family of keratins, fibrous structural proteins of nails, hair, and epidermis, keratin types 15 and 19 are expressed in the cells of the bulge of the hair follicle that are suspected to be stem cells.[18,19] Some data also indicate that the nuclear expression of p63, a transcription factor of the tumor suppressor p53 gene family, seems to be connected with the stemness of these cells; however, other data suggest that nuclear expression of this molecule is more related to the beginning of the program of epithelium stratification.[20] CD34, known as a marker of hematopoietic cell lineage, was also detected in the cells of the bulge region of the hair follicle.[21] Finally, a specific glycosylation of proteins also represents a useful marker of EpSC differentiation. Labeled plant and animal lectins (glycoproteins that specifically bind carbohydrate structures) can be used for functional analysis of carbohydrate epitopes of epidermal cells. In human and porcine models, basal cells, including EpSCs, expressed the animal lectin galectin-1, or they expressed epitopes reactive to this galectin.[22–24] On the other hand, more differentiated epidermal cells in the suprabasal layers were positive for animal lectin galectin-3-reactive epitopes[22] and also for plant lectin *Dolichos biflorus* agglutinin-reactive epitopes.[25]

EpSCs are multipotent and may differentiate into various types of epidermal cells, such as keratinocytes, hair cells, and sebaceous gland cells. EpSCs transfected with GFP and injected into a blastocyst can be detected later in many tissues of the embryo, indicating the potential of these cells to differentiate into various cell types.[26] Interestingly, epidermal cells that are harvested from porcine hair follicles form spheroid cell clusters *in vitro* that are similar to neurospheres or embryonic bodies.[24] Highly multipotent stem cells (probably of neural crest origin) were also isolated from porcine and mouse skin[27,28] and from the bulge region of the outer root sheath of the hair follicle.[29] The possible relationship between these neural crest stem cells and the epidermal stem cells is not clear.

The maintenance of stemness in epidermal cells cultured *in vitro* is problematic and therefore an *in vivo* model is needed. The accessibility of human tissue is highly limited and excludes the use of metabolic experiments such as the application of labeled precursors of DNA. Therefore, animal models are extensively used. To date, most of the progress in EpSC research has been made in mice. The main advantage of the mouse model is the possibility of preparing genetically manipulated animals (transgenics or knockouts) that can be produced inexpensively, owing to the short generation time in small rodents. Unfortunately, the murine epidermis presents substantial morphological differences when compared to human tissue. In contrast, the remarkable similarity of human and pig epidermis was noted many years ago, and thin, dermoepidermal porcine grafts are even being used as a provisional cover of skin defects in clinical practice. The economy of the work is also important, because the samples of porcine epidermis can be easily obtained from a local slaughterhouse and used for *in situ* and *in vitro* experiments. From a methodologic point of view, the porcine hair follicles, including their sheaths, can be easily microdissected and used for cultivation experiments, and most antibodies against human epidermal markers crossreact with porcine antigens. These data presented here indicate that porcine epidermis has an important role as a model in epidermal stem cell research and it cannot be excluded from wound healing research.

NEURAL STEM CELLS

In recent years, the interest in neural development, plasticity, and regeneration has increased tremendously, and a wide variety of strategies and techniques has been developed and applied in this scientific field. This includes neurotransplantation, which has become one of the modern strategies of choice in both basic and applied neuroscience. In the early 1990s, neural stem cells (NSCs) from different stages of central nervous system (CNS) development were discovered, and their great plasticity and multipotency were revealed. Operationally, these cells are defined by their ability to self-renew and to differentiate into cells of all glial and neuronal lineages throughout the neuraxis.[30] Today, the possibility of maintaining NSCs in culture, and to expand and manipulate them quickly, makes them a popular source of grafting material in the biomedical research community, having a great therapeutic potential for the treatment of CNS injuries and neurodegenerative diseases.

Multipotent NSCs can be isolated from the neurogenic zones of developing or adult CNS and grown *in vitro* under well-defined conditions, either as monolayer cultures or as floating multicellular aggregates, called "neurospheres." Although up

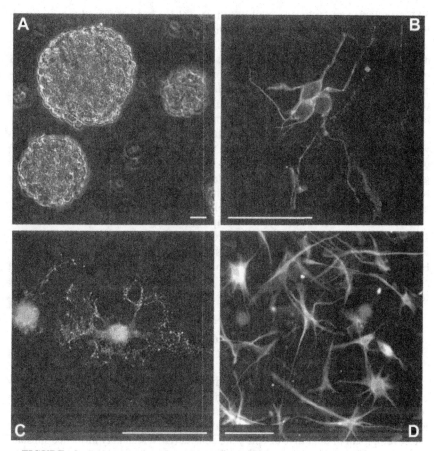

FIGURE 2. Examples of cultured NSCs isolated from miniature pig brain. Neural progenitor cells from the subventricular zone of E40 porcine fetuses can be grown *in vitro* similarly to human NSCs and differentiated into neurons, astroglia, and oligodendrocytes. (**A**) Phase-contrast image of porcine NSCs after 10 days *in vitro* (10DIV): the cells form neurospheres when cultured in serum-free medium supplemented with EGF and bFGF. After plating on laminin-coated coverslips and 12 DIV without growth factors, cells expressing the neuronal marker β-III-tubulin (**B**), the oligodendroglial marker CNPase (**C**), and the astroglial marker GFAP (**D**) can be detected. Scale bars represent 20 μm.

to now, no markers specific for NSCs have been established, these cells can be defined by the absence of more mature neural markers such as β-III tubulin (TuJ-1), glial fibrillary acidic protein (GFAP), or the neuronal nuclear marker NeuN, and by the presence of several markers for undifferentiated cells, including the intermediate filament protein nestin, the RNA-binding protein Musashi-1, and transcription factors such as Sox-1[31], Otx-1, Otx-2, NeuroD1, and neurogenin-2.[32]

For proliferation and survival *in vitro*, NSCs require mitogens such as basic fibroblast growth factor (bFGF) and epidermal growth factor (EGF). For human NSC cultures, which are more delicate and difficult to work with than, for example, rodent

NSCs, inclusion of leukocyte-inhibitory factor (LIF) has been used together with EGF to prolong their *in vitro* lifespan.[33] Differentiation of NSCs can be triggered by the removal of growth factors and plating on poly-L-lysine, laminin, or other standard culture growth surfaces. There exists regional specificity in the differentiated progenies of NSCs; for example, spinal cord NSCs produce only spinal cord progeny,[34] and only cells isolated from the midbrain consistently differentiate into functional dopaminergic neurons.[35] Such differentiation programs, at least *in vitro,* can be influenced by the addition of various growth factors and cytokines (FIG. 2).

Recently, we compared the *in vitro* behavior of porcine NSCs isolated from whole brain (embryonic day 25, E25) and the neocortical subventricular zone (E40) of miniature pig fetuses with human NSCs isolated from 10-week-old fetuses (a kind gift from Clive Svendsen, University of Wisconsin). Intriguingly, both cell types can be grown under the same growth conditions and behave comparably with respect to growth factor requirements and patterns of differentiation (glia/neuronal index, cell type profiles, etc.). Both cell types display similar migration speeds and relatively slow growth rates, and require a similar frequency of careful cell culture passaging, which must avoid trypsinization. In this sense, the pig and human NSCs appear to be more closely related to each other than either of them is to mouse NSCs, which suggests that the behavior of porcine NSCs could well be used to predict and study that of human NSCs.

THE MINIATURE PIG AS A MODEL FOR NEURODEGENERATIVE DISEASES

Animal models of neurodegenerative diseases are important in current biomedical research that is exploring the therapeutic use of NSCs. In these models, by far the most results come from studies on rodents, particularly mice. Unfortunately, rodent and primate brains present several important differences when size, complexity, and the development of cognitive circuitry are considered, which may influence substantially the behavior of implanted NSCs. Therefore, before any clinical trials can be envisaged based on results from small-animal models, the experiments need to be validated in species related as closely as possible to man. The closest one can get today to the human situation in an animal model using NSC transplantation is to graft human NSCs into a non-human primate.[36] Although the non-human primate model seems ideal, the technical and organizational difficulties and high costs related to it beg for an alternative species. Therefore, we need animal models that reflect as closely as possible the geometry and complexity of the cytoarchitecture found in the human CNS, yet would still be more affordable than primates.

Miniature pigs have a relatively large brain with a human-like blood supply and immunologic response characteristics. Together with the similarities described above between cultured porcine and human NSCs, which distinguish both from their mouse counterparts, miniature pigs and grafting of NSCs isolated from their fetal brains (allografting) could thus represent a unique and viable alternative to experiments addressing questions related to the behavior of NSCs in primates.

Recently, fetal porcine neural tissue was used in several xenotransplantation studies in rodent models of neurodegenerative disorders[37] and also in several phase 1 clinical trials in patients with Parkinson's or Huntington's disease.[38] *In vitro* expanded and modified fetal porcine neural progenitor cells were also used for re-myelination experiments in adult rodent CNS.[39] However, pigs can serve not only as a source of cells for xenotransplantation. With the development of transgenic and chemically induced models of neurodegenerative disorders, pigs also represent a valuable model for allogenic transplantation experiments.[40] Their CNS size and physiology and their longer lifespan make swine also suitable model animals for long-term evaluation of the safety and efficacy of possible cell therapies for CNS disorders.

CONCLUDING REMARKS

Rapid progress in basic biomedical research, mostly conducted on small laboratory rodents, has generated a huge amount of experimental data. However, before this newly gained knowledge can find its way into designs of new therapies, we need to validate it on animal models more closely related to humans. On the basis of selection of examples adduced in this paper, we are convinced that the laboratory miniature pig can represent such a model. Compared to non-human primates, the primary candidate species for bridging this gap, pigs are cheaper and easier to maintain in controlled conditions. Their human-like physiology assures a high relevance of the data obtained in this species for human-related therapeutic research. Moreover, the recent advances in methods of transgenic animal production will allow a certain "humanization" of pigs for use in xenotransplantation, and also the creation of transgenic disease models, complementing rodent ones. With the miniature pig gestation period of 114 days and litters up to 12 piglets, enough experimental animals can be obtained, and a lifespan of 12–18 years allows long-term experiments evaluating the safety and efficacy of possible therapies. As we have described in this text, miniature pigs are already extensively used in several fields of biomedical research, and we firmly believe that the importance of these animals as a biomedical model will increase even further in the near future.

ACKNOWLEDGMENTS

This work was supported by the Czech Ministry of Education Grant LN00 A0065 and an Iowa State University Special Research Initiation Grant.

REFERENCES

1. ROTHSCHILD, M.F. & A. RUVINSKY. 1998. Genetics of the Pig. CAB International. Wallingford, UK.
2. HASSOLD, T. & P. HUNT. 2001. To err (meiotically) is human: the genesis of human aneuploidy. Nat. Rev. Genet. **2:** 280–291.
3. ELLEDEROVÁ, Z. *et al.* 2004. Protein patterns of pig oocytes during in vitro maturation. Biol. Reprod. **71:** 1533–1539.
4. RUI, R. *et al.* 2004. Attempts to enhance production of porcine chimeras from embryonic germ cells and preimplantation embryos. Theriogenology. **61:** 1225–1235.
5. UCHIDA, M. *et al.* 2001. Production of transgenic miniature pigs by pronuclear microinjection. Transgen. Res. **10:** 577–582.

6. MOFFAT, A.S. 1998. Improving gene transfer into livestock. Science **282:** 1619–1620.
7. LAVITRANO, M. *et al.* 2002. Efficient production by sperm-mediated gene transfer of human decay accelerating factor (hDAF) transgenic pigs for xenotransplantation. Proc. Natl. Acad. Sci. USA **99:** 14230–14235.
8. JAENISCH, R. 1976. Germ line integration and Mendelian transmission of the exogenous Moloney leukemia virus. Proc. Natl. Acad. Sci. USA **73:** 1260–1264.
9. CHAN, A.W. *et al.* 1998. Transgenic cattle produced by reverse-transcribed gene transfer in oocytes. Proc. Natl. Acad. Sci. USA **95:** 14028–14033.
10. PFEIFER, A. *et al.* 2002. Transgenesis by lentiviral vectors: lack of gene silencing in mammalian embryonic stem cells and preimplantation embryos. Proc. Natl. Acad. Sci. USA **99:** 2140–2145.
11. WOLFGANG, M..J. *et al.* 2001. Rhesus monkey placental transgene expression after lentiviral gene transfer into preimplantation embryos. Proc. Natl. Acad. Sci. USA **98:** 10728–10732.
12. HOFMANN, A. *et al.* 2003. Efficient transgenesis in farm animals by lentiviral vectors. EMBO Rep. **4:** 1054–1060.
13. LAI, L. *et al.* 2002. Production of alpha-1,3-galactosyltransferase knockout pigs by nuclear transfer cloning. Science **295:** 1089–1092.
14. LOI, P., J. FULKA, JR. & G. PTAK. 2003. Amphibian and mammal somatic-cell cloning: different species, common results? Trends Biotechnol. **21:** 471–473.
15. BICKENBACH, J.R. 2004. Isolation, characterization, and culture of epithelial stem cells. *In* Methods in Molecular Biology. Vol. **289:** 97–102. Humana Press. Totowa, NJ.
16. WATT, F.M. 2002. The stem cell compartment in human interfollicular epidermis. J. Dermatol. Sci. **28:** 173–180.
17. KAUR, P. & A. LI. 2000. Adhesive properties of human basal epidermal cells: an analysis of keratinocyte stem cells, transit amplifying cells, and postmitotic differentiating cells. J. Invest. Dermatol. **114:** 413–420.
18. LAROUCHE, D. 2004. Keratin 19 as a stem cell marker in vivo and in vitro. *In* Methods in Molecular Biology. Vol. **289:** 103–110. Human Press, Totowa, NJ, USA.
19. LIU, Y. *et al.* 2003. Keratin 15 promoter targets putative epithelial stem cells in the hair follicle bulge. J. Invest. Dermatol. **121:** 963–968.
20. KOSTER, M. I. 2004. p63 is the molecular switch for initiation of an epithelial stratification program. Genes Dev. **18:** 126–131.
21. TUMBAR, T. 2004. Defining the epithelial stem cell niche in skin. Science **303:** 359–563.
22. CHOVANEC, M. *et al.* 2004. Decrease of nuclear reactivity to growth-regulatory galectin-1 in senescent human keratinocytes and detection of non-uniform staining profile alterations upon prolonged culture for galectins-1 and -3. Anat. Histol. Embryol. **33:** 348–354.
23. PURKRÁBKOVÁ, T. *et al.* 2003. New aspects of galectin functionality in nuclei of cultured bone marrow stromal and epidermal cells: biotinylated galectins as tool to detect specific binding sites. Biol. Cell **95:** 535–545.
24. KLÍMA, J. *et al.* 2005. Comparative phenotypic characterization of keratinocytes originating from hair follicles. J. Mol. Histol. **36:** 89–96.
25. DVOŘÁNKOVÁ, B. *et al.* 2002. *Dolichos biflorus* agglutinin-binding site expression in basal keratinocytes is associated with cell differentiation. Biol. Cell **94:** 365–373.
26. LIANG, L. *et al.* 2004. As epidermal stem cells age they do not substantially change their characteristics. J. Invest. Dermatol. Symp. Proc. **9:** 229–237.
27. FERNANDES, K.J.L. *et al.* 2004. A dermal niche for multipotent adult skin-derived precursor cells. Nature Cell Biol. **6:** 1082–1093.
28. DYCE, P.W. *et al.* 2004. Stem cells with multilineage potential derived from porcine skin. Biochem. Biophys. Res. Commun. **316:** 651–658.
29. SIEBER-BLUM, M. *et al.* 2004. Pluripotent neural crest stem cells in the adult hair follicle. Dev. Dynamics **231:** 258–269.
30. BAIZABAL, J.M. *et al.* 2003. Neural stem cells in development and regenerative medicine. Arch. Med. Res. **34:** 572–588.
31. AUBERT, J. *et al.* 2003. Screening for mammalian neural genes via fluorescence-activated cell sorter purification of neural precursors from Sox1-gfp knockin mice. Proc. Natl. Acad. Sci. USA **100** (Suppl. 1): 11836–11841.

32. CAI, J. *et al.* 2003. Identifying and tracking neural stem cells. Blood Cells Mol. Dis. **31:** 18–27.
33. WRIGHT, L.S. *et al.* 2003. Gene expression in human neural stem cells: effects of leukemia inhibitory factor. J. Neurochem. **86:** 179–195.
34. WALDER, S. & P. FERRETTI. 2004. Distinct neural precursors in the developing human spinal cord. Int. J. Dev. Biol. **48:** 671–674.
35. STORCH, A. *et al.* 2001. Long-term proliferation and dopaminergic differentiation of human mesencephalic neural precursor cells. Exp. Neurol. **170:** 317–325.
36. OUREDNIK, V. *et al.* 2001. Segregation of human neural stem cells in the developing primate forebrain. Science **293:** 1820–1824.
37. ARMSTRONG, R.J. *et al.* 2003. Transplantation of expanded neural precursor cells from the developing pig ventral mesencephalon in a rat model of Parkinson's disease. Exp. Brain. Res. **151:** 204–217.
38. FINK, J.S. *et al.* 2000. Porcine xenografts in Parkinson's disease and Huntington's disease patients: preliminary results. Cell Transplant. **9:** 273–278.
39. SMITH, P.M. & W.F. BLAKEMORE. 2000. Porcine neural progenitors require commitment to the oligodendrocyte lineage prior to transplantation in order to achieve significant remyelination of demyelinated lesions in the adult CNS. Eur. J. Neurosci. **12:** 2414–2424.
40. DALL, A.M. *et al.* 2002. Quantitative [^{18}F]fluorodopa/PET and histology of fetal mesencephalic dopaminergic grafts to the striatum of MPTP-poisoned minipigs. Cell Transplant. **11:** 733–746.

Graft/Host Relationships in the Developing and Regenerating CNS of Mammals

VÁCLAV OUREDNIK AND JITKA OUREDNIK

Department of Biomedical Sciences, College of Veterinary Medicine,
Iowa State University, Ames, Iowa 50011, USA

ABSTRACT: A new light was shed on the utility of neural grafts when it was recognized that donor tissues and cells offer more than a source of immature progenitors potentially capable of cell replacement: First, they have the inherent capacity to produce multiple trophic and tropic factors promoting cell survival and tissue plasticity often characteristic of the immature central nervous system (CNS). Second, by their interaction with the host microenvironment via cell/cell and cell/ECM interactions, these grafts are capable of re-establishing homeostasis, which can be, for example, reflected in rescue and protection of host elements from harmful influences. This second capacity of donor cells relies, in part, also on a "dormant" but still present regenerative capacity of mature or even aged CNS and on the possibility of its mobilization in the damaged nervous system by neural grafts. For this to occur efficiently after transplantation, a bi-directional dialogue between donor and host cells must gradually be established, in which both "partners" transmit signals (cell/cell contact, molecular messengers), "listen to" and "understand" each other and are able to react by modifying their own plasticity- and development-related programs. Thus, for the best possible recovery of functionality in the injured adult and aged nervous system, neurotransplantation must always try to find optimal conditions for all three of the mentioned qualities of neural grafts, especially for the protection and/or reactivation of neural circuitry embedded in non-neurogenic CNS areas. Once fully understood, this newly recognized aspect of neurotransplantation (and topic of this review) might, someday, even allow the recovery of systems that would otherwise be doomed, such as cognition- and experience-related circuitry.

KEYWORDS: CNS; regeneration; neuroprotection; rescue; graft; neurotransplantation; neural stem cell

GRAFTING AS A TOOL FOR STUDYING CNS DEVELOPMENT AND REGENERATION

Experimental neurotransplantation was first attempted more than one hundred years ago and used by neuroanatomists and embryologists as a tool for addressing basic questions in early ontogeny. While grafting into the mammalian nervous sys-

Address for correspondence: Dr. Václav Ourednik, Department of Biomedical Sciences, College of Veterinary Medicine, Iowa State University, Ames, IA 50011. Voice: +1-515-294-3719; fax: +1-515-294-6449.

voured@iastate.edu

Ann. N.Y. Acad. Sci. 1049: 172–184 (2005). © 2005 New York Academy of Sciences.
doi: 10.1196/annals.1334.016

tem (NS) remained for a long time unsuccessful, transplantation in other species, such as fish, amphibians, reptiles, and birds, was relatively easy and could be done in the very early stages of embryonic development—during the incubation of the egg. These experiments helped to resolve some fundamental problems of developmental biology and led to essential discoveries such as neuronal induction, cephalization, and body polarization. It became evident that the ontogeny of individual organs depends also on the development of surrounding tissue, and that close communication between individual cells and structures is needed and regulated according to a genetically defined developmental program in time and space.

In contrast to lower vertebrates and birds, the mammalian class is characterized by substantially diminished plasticity in the adult NS. This is mainly due to the highly specific and complex neural connectivity, characterized by precisely controlled communication between billions of cell types with defined tasks and characteristics. Especially in higher mammals such as primates, many of these connections are of a cognitive nature and are pruned and specified only postnatally, obeying learning- and experience-dependent input and escaping early development-based repair mechanisms. It has also become apparent that the number of precursor cells and, importantly, their potential to differentiate into various types of neural cells, decreases progressively throughout the development of an individual. The regression of germinative and neurogenic zones is probably one of the main reasons for the progressive loss of the nervous system's restorative capacity.[1] Other inhibitory factors are continuously being discovered in the molecular composition of the adult CNS environment[2] and in the formation of the blood–brain barrier, which prevents the entrance not only of many noxious substances, but also of endogenous and systemically applied therapeutic substances into the CNS from the vascular compartment.

Thus, it seems to be almost impossible that the cytoarchitectonic patterns established during the developmental and learning phases of the CNS by genetically coded spatio-temporal morphogenetic gradients, can ever be re-established when damaged. However, encouraging evidence exists for anatomical and functional modification and adaptation of connections during normal life as well as in the injured adult.

The ancient concept that Arnemann formulated in 1787 rejected, for decades, any regenerative possibility in the postnatal NS of mammals, both peripheral and central.[3] Only in 1901, did Bielchowski and Cajal observe that a peripheral nerve can regenerate when it stays in contact with its cell body in the CNS.[4] With the aim to "transfer regenerative propriety of the peripheral nervous system into the central nervous system," Ramon y Cajal transplanted peripheral elements (peripheral nerves and dorsal root ganglia) into the brain.[5] From these experiments, he postulated that the determination of axonal regeneration depends more on the microenvironment of the peripheral axon than on the cell body situated within the CNS.

Today, a new concept of the interaction of intrinsic and extrinsic developmental factors is dawning and opening new avenues. It is of prime interest for modern neuroscience research to unravel the molecular and functional principles underlying the development and regeneration of CNS cytoarchitecture and the character and roles of the ontogenic components. To what extent these principles are determined by genetic and epigenetic (other neurons, neuroglial cells, chemical gradients, learning, disease, and other factors) means, and how they mold the structure and function of neurons and glial cells, still remains largely unknown.

Thus, current neurotransplantation research not only offers hope for human medicine, but has also become an important tool for gaining new insights into neural plasticity, that is, into dynamic mechanisms at all levels of the neuroaxis, such as axonal reorganization and synaptogenetic reactivity. In a broader biological sense, all of these activities are responsible for the structural and functional establishment and agility of the mammalian NS. Especially in the brain, the cytoarchitecture is continuously sculpted in immediate response to experience and learning processes, based on external input and the interaction of the individual with the environment. This, however, causes questions about determinism and indeterminism, or nature versus nurture, to gain renewed importance in neuroscience, and begs for new experimental approaches. Such questions can be investigated in two classic ways: either by observing cellular interrelations in normal development or by observing patterns of variation and/or invariance which may arise in abnormal circumstances as a result of particular mutations or chemical or mechanical insults. In this second approach, neurotransplantation has joined in as a strategy to explore CNS development and plasticity in non-traditional ways. It soon became one of the modern strategies of choice in both basic and applied neuroscience and today helps answer questions concerning cell proliferation, differentiation, programmed cell death, and target-oriented migration in the CNS, as well as the formation of neural circuitry, CNS immune response, and many other issues of critical interest.

TIMID BEGINNINGS

The fact that in some animals, and during early development, regeneration of the CNS does occur soon tempted researchers to engage in neurotransplantation experiments. The first attempts occurred as early as 1890, followed by others at the beginning of the 20th century. For example, Thompson transferred allografts (transfer of tissue between individuals within one species) and xenografts (between individuals belonging to different species) into the cerebral cortex of a dog,[6] and Saltykow performed allografting between adult rabbits.[7] Since, at that time, there existed little knowledge about the immune response and its causes, these pioneer experiments with grafts into the adult CNS within or between species did not lead to long-term survival of the transplanted material. The repeated failure led only to skepticism about mammalian neurotransplantation and was taken as a confirmation of the plastic rigidity of the mammalian CNS.

The modern concept of neurotransplantation in mammals did not appear until 1917, when E.H. Dunn transferred embryonic neural tissue into the rat CNS, thereby formulating two fundamental principles: (1) the graft has to be composed of immature tissue, and (2) it must get trophic support from the recipient.[8] This original approach was thus to take advantage of the highly plastic characteristics of dividing cells residing in the germinative regions of the developing CNS. The vivacity and adaptability of the immature neural tissue contrasted sharply with the regenerative rigidity of the adult mammalian CNS, and the possibility that immature donor tissue might deliver new "building matter" or missing substances into the impaired host CNS soon sparked new hopes. Nevertheless, despite all the promises, this innovative "avant-garde" concept of neurotransplantation introduced by Dunn was abandoned for many decades to come.

A new wave of neurotransplantation finally started to take off in the late 1970s, when Perlow and collaborators demonstrated in rodents that non-differentiated dopaminergic (DA) neurons could compensate for a motor deficit induced by the unilateral damage of the substantia nigra (SN).[9] From that time on, experiments employing grafting into mammals multiplied and diversified tremendously, rising from a purely empirical phase into a stage firmly rooted in the rational bases of anatomy, physiology, and molecular biology. The quickly accumulating knowledge prompted the development of various animal models recapitulating certain neurological syndromes in man and allowed theoretical and some practical investigations envisaging the application of neurotransplantation in clinical therapy.

A RENAISSANCE OF TRANSPLANTATION INTO THE CNS

The appeal of clinical relevance of neural grafts became particularly strong during the late 1980s, when Dunn's original approach was again remembered, and researchers started to experiment with neural grafts dissected from different parts of the fetal or early postnatal CNS. This approach took advantage of the highly plastic characteristics of dividing cells residing in the germinative regions of the developing CNS and, simultaneously, of the relatively higher immunological tolerance of the CNS (for a long time, it was believed that the CNS is an "immunologically privileged" organ in the body, but this opinion was later substantially revised (see, for example, Xiao and Link[10]). The so-called "primary grafts" were transferred, in solid form or as cell suspensions, into intact pre- and postnatal brains and/or into the CNS of animals serving as models of various neurological disorders in humans. Particularly in the treatment of Parkinson's and Huntington's diseases, this approach eventually resulted in impressive anatomical and biochemical repair, paralleled by significant amelioration of the behavioral symptoms.[11-15]

Although primary neural grafts do occupy a firm place in the history of neurotransplantation, the limitation of material accessibility, concerns about its purity and viability, as well as political and ethical issues led many researchers to look for alternatives. These appeared in the early '90s in form of the freshly discovered neural stem cells (NSCs), demonstrating great plasticity and multipotency (see, for example, Refs. 16 and 17). The possibility of maintaining NSCs in culture, and expanding and manipulating them quickly and efficiently, stirred great interest in the biomedical research community as representing a potentially rich source of grafting material to replace fetal tissues.

Soon, the applications of stem cells in transplantation became as diversified as their sources and characteristics, which can be divided conceptually into various groups according specific parameters. While NSCs are isolated directly from the CNS (fetal germinative zones or adult neurogenic niches), other cells with stem cell properties can readily be obtained from various non-neural organs and/or in the form of pluripotent embryonic stem cells from very early embryos prior to germ-layer formation (blastocysts).[18] The source of NSCs is, however, of crucial importance and defines their biology, their ability to survive, migrate, and incorporate into the host

CNS, as well as their potential to differentiate into neural cells. These and other properties of NSCs may be further modified experimentally with the aim of enforceing the cells' capability to serve as vectors for the delivery of defined substances, to enhance their migration, or to guide their differentiation predominantly into one specific cell type.[19]

Depending on the donor material, neural grafts (both tissue and NSCs) have been proposed to help morphological and functional recovery of the recipient CNS by adopting one or more of the following strategies: (1) replacing affected cell populations (or structural components like myelin) and their connections, which would correlate with functional recovery; (2) delivering missing neuroactive molecules, such as enzymes and neurotransmitters, by targeted genetic engineering of the grafted cells, as occurs, for example, in most of the current therapeutic approaches for Parkinson's disease; (3) creation of highly growth-permissive tissue "bridges" for host axonal regeneration and target-oriented guidance of growing axons; and (4) provision of neurotrophic and other regeneration-promoting substances through natural and endogenous secretion by the transferred cells supporting the survival and growth of graft and host neural structures (this last strategy is discussed in more detail below).

The application of these principles in neurotransplantation has provided us with a wealth of information about normal CNS development and the characteristics of a disorder-dependent variability of the host environment and its impact on the behavior of the grafted tissues and cells. However, it also constantly highlights our persistent lack of understanding of the mechanisms controlling stem cell differentiation under the host's microenvironmental influence. Especially in the adult and aged CNS, the prevailing non-neurogenic and growth-inhibitory milieu has a strong negative impact on that process[20–22] and substantially reduces the success rate of neurotransplantation.

To unravel at least some of the mysteries pertaining to the problems alluded to in the last paragraph, we need to realize that in neurotransplantation, it is not only the graft that is exposed to and that reacts to environmental changes nor is it only the recipient that secretes signaling molecules influencing the behavior of donor cells. After grafting, an intimate cellular and molecular relationship is established between graft and host in which both try to reach an "ideal" homeostatic equilibrium. In this fusion of two worlds of quite diverse characteristics, only a detailed knowledge of the laws that govern the resulting reciprocal interactions will allow us to take advantage of them for clinical benefits.

GRAFT-INDUCED CNS PLASTICITY

Today, it seems more and more obvious (and we were among the first proponents of this idea back in the early '90s[23]) that the ability to replace damaged or missing cell populations or to deliver missing enzymes and structural molecules is by far not the only beneficial feature of grafted fetal tissue and NSCs that deserves all the credit. In the light of new findings regarding postnatal CNS plasticity[24] and intercellular graft/host communication,[25] a much broader image begins to emerge of the roles and consequences of this complex cellular/molecular dialogue between essentially two different worlds coming together. It forces us to go back and to try to understand regeneration of the grafted CNS in light of mechanisms and principles governing CNS development in order to fully apprehend the full spectrum of events that occurs during the intricate dialogue between graft and host cells.

In the early '90s, together with the late Hendrik Van der Loos, we worked on repairing a mechanically damaged somatosensory cortex by transplanting fetal tissue into juvenile mice. After a series of grafting experiments, we observed that, instead of replacing the damaged and missing cells in the unilaterally created cortical cavity in the primary somatosensory barrel field area, primary neural tissue isolated from embryonic day 14 (ED14) mouse neopallium could induce repair from within the juvenile host in a manner not seen spontaneously in the absence of transplantation.[23] The reconstituted cavity (in 70–80% of cases) appeared to become "filled" by relatively well-organized cortical tissue of *host* origin that even presented cellular arrangements reminiscent of barrels, which even displayed, at least partially, electrical activity (unpublished observations). The broader significance and applicability of this graft-induced but host-dependent cortical reconstruction was later confirmed in a xenotransplantation paradigm in which mouse tissue was transplanted into the lesioned CNS of kittens.[26]

In an attempt to ascertain a mechanism underlying this graft-evoked regeneration, the possibility of *de novo* neurogenesis was investigated in animals pulse-labeled with tritiated thymidine ($[^3H]T$) or bromo-deoxyuridine (BrdU), both nucleotide analogues labeling dividing cells during the S phase of proliferation. Our hopes that we could reactivate the formation of neurons in the host were, however, not met. The vast majority of $[^3H]T$- and BrdU-labeled host cells turned out to be of glial and endothelial origin (although no significant gliosis or glial scar were ever found). The formation of new host neurons, on the other hand, was quite rare, and therefore insufficient to account for the substantial tissue recovery. Thus, alternative mechanisms needed to be explored that could have led to the observed plasticity. Such mechanisms may include remodeling of surrounding tissue by existing postmitotic cells, activation and differentiation of dormant progenitor cells present in the postmitotic CNS, and prevention of necrosis and secondary host cell death in the vicinity of the cavity.

At that time, on the basis of the observation of graft-induced host plasticity in damaged mouse and kitten neocortex, we predicted that this "regenerative phenomenon" is likely to depend on a favorable cellular and molecular environment that had to be created in a specific and reciprocal communication between graft and host elements, since neither non-grafted cavities nor cavities receiving non-neural fetal tissue presented any signs of recovery. Certain cellular elements within the fairly heterogeneous fetal tissue grafts seemed to modify the brain environment such that intrinsic, albeit latent, regenerative responses by the host were being uncovered, triggered, and amplified. The long-range effects of grafted tissue in some of these experiments suggested at least a partial role for secreted and diffusible substances mediating the necessary graft/host interaction. This assumption was corroborated by others investigating the production of growth factors by grafts and the responsiveness of the adult CNS to them.[27,28] Our thoughts led us to speculate that a source of such trophic and neuroprotective factors might well be the non-differentiated stem cells residing in the germinal parts of our neopallial grafts (which always included parts of the fetal ventricular zone).

COOPERATION OF HOST AND DONOR STEM CELLS IN THE
DEVELOPMENT OF GRAFTED FETAL CNS

In both the developing and the adult nervous system the behavior of cell precursors is regulated by environmental factors encountered either in the germinative zones or in the stem cell niches. Thus, the ultimate NSC behavior *in vivo* is the final cellular response to converging signals coming from various neighboring cell types such as endothelial and glial cells, or from the local extracellular matrix (ECM).[29,30] This continuing metamorphosis is reflected in both genetic and epigenetic components of ontogeny.

What happens, however, when such continuity of the intact CNS development is affected by transplantation—the coming together of two worlds of quite different cellular and biochemical properties? Today, we know that the fate of *both* interacting entities can begin to change and to be redefined depending on the character and intensity of mutual signals and the presence of cells capable of receiving and reacting to them. Thus, in a graft, represented by primary fetal tissue or by certain types of stem cells (SCs), the molecular principles underlying its response to the environment are governed by the developmental stage and the area from which its material was derived, and by the way it was processed prior to transplantation. Although this mutual signaling and responding is still far from being completely understood, it seems that the underlying "language" is of a universal character and can be studied even after xenografting.

To address some of the issues discussed in the preceding paragraph, we recently inspected the behavior of human NSCs grafted *in utero* into the ventricles of a mid-gestation non-human primate brain.[31] We observed that the donor cells not only could survive the xenotransplantation, but they also became integrated into the organogenetic program of the host. The injected cells entered the subventricular zone, interdigitated with the endogenous NSCs, and both together subdivided into different cell components of the developing brain. This "teamwork" of donor and host NSCs was reflected in a prevalent morphogenetic scenario. Thus, many of the NSCs soon left the subventricular zone, migrated along the processes of radial glia into the growing brain parenchyma, and started to differentiate into various types of neural cells, while contributing to the spatio-temporal morphogenetic gradient of the fetal monkey brain. Other subpopulations of donor and host NSCs remained undifferentiated and formed quiescent pools of multipotent cells, possibly for later use during ontogeny or maintenance of CNS homeostasis.

The presence of dormant undifferentiated cells within the adult CNS has by now been firmly established, and we believe that these cells may be used in later life for self-repair and literal cell turnover or to exert a protective effect upon imperiled juxtaposed neurons by providing trophic support.[27,28] The idea that this "chaperone" population can be augmented by transplantation and used effectively for the rescue of impaired brain populations[25] is effectively demonstrated in the experiments described in the next chapter.

NSCs RESCUE DYSFUNCTIONAL NEURONS

The slow disintegration of neural function associated with aging and many neurodegenerative diseases provides an ideal opportunity to test the putative NSC capacity to rescue dysfunctional cells in an environment *not* confounded by excessive necrosis, excitotoxicity, anoxia, or trauma.

One well-characterized and easily-identifiable neuronal cell type with stereotypical projections that is typically compromised in the aged brain—the dopaminergic (DA) neuron—was used for the following study.[25] To accelerate the process of aging and to simulate more of the pathological processes known to accompany neurodegenerative diseases such as Parkinson's disease, aged mice were exposed to systemic injections of MPTP (1-methyl-4-phenyl-1,2,3,6-tetrahydropyridine), a toxin selectively affecting DA neurons. The animal model was carefully designed with the aim of not killing the mesencephalic DA neurons, but only to cripple them chronically in both cerebral hemispheres. Unilateral implantation of murine NSCs above the right ventral tegmental area (VTA) occurred 1 and 4 weeks post MPTP. In contrast to young or intact aged brains, the progressive immigration of the donor cells into both hemispheres was very extensive in the aged, MPTP-treated hosts. While in sham-transplanted animals (receiving MPTP treatment but no NSC graft) no recovery ever happened, the neural transplants resulted in a gradual reconstitution of the damaged mesostriatal system; first uni- and then bilateral, the recovery became apparent through the re-expression of two of its key markers, tyrosine hydroxylase (TH) and the membrane-bound DA transporter (DAT). Interestingly, the steady reactivation gradient of TH and DAT expression in the DA pathway was associated with corresponding D-amphetamine-evoked rotation of the animals during the first week after grafting. This rotation was a sign of an initial post-transplantation imbalance in dopamine levels between the two brain hemispheres and a consequence of the progressive recovery gradient in the DA system. Although the restoration of DA function was graft-dependent, it was not predominantly due to differentiation of the donor NSCs into DA neurons. While there was a sporadic and spontaneous conversion of NSCs to DA cells contributing to nigral reconstitution in the depleted mesencephalic areas, the vast majority of the recovered mesencephalic neurons were actually *"rescued"* host cells. Thus, in this example, a damaged host CNS did benefit from transplanted NSCs not because of their capacity to replace cells, but mainly because of a channel of communication with host cells leading to significant sparing and rescue of damaged but still persisting host cytoarchitecture.

This often ignored new possibility of saving damaged host circuitry and increasing the regenerative capacity in an adult injured CNS by a plasticity-promoting dialogue between graft and host has been, by now, corroborated in several other studies. Thus, it was demonstrated that NSCs seeded upon a synthetic biodegradable scaffold and grafted into the hemi-sectioned adult rat spinal cord, induced a significant improvement in animal movement by reduction of necrosis in the surrounding parenchyma and prevention of extensive secondary cell loss, inflammation, and formation of a glial scar. Here again, as tract tracing and GAP-43 immunoreactivity showed, the host tissue displayed regenerated neurites *not* derived from donor NSCs, but of recipient origin.[32] A more recent study in the lesioned spinal cord showed convincingly how such host neurite sprouting can be influenced by variation of the spectrum of donor cell–released trophic factors which can, for example, selectively stimulate the regeneration of either motor or sensory fibers.[28] Finally, a substantial reconstitution of the brain parenchyma and structural connectivity was reported when NSCs were transferred into regions of extensive brain degeneration caused by hypoxia, particularly when donor cells were transiently supported by biodegradable scaffolds.[33] Also, in this case, the injured brain interacted reciprocally with the exogenous NSCs, which resulted in partial tissue reconstitution and host neurite sprouting.

The interaction between neural graft and host CNS is very complex and is likely to involve not only trophic and growth factors, but also cell-recognition molecules and the ECM.[29,30,34] The latter two, in particular, have so far hardly been explored in the context of neurotransplantation. To explore their involvement in the engagement of NSCs in regeneration, in the following study, we therefore elected to examine as a prototypical neural adhesion molecule, one of the earliest and best characterized of that class of molecules, the integral membrane glycoprotein L1 (Ourednik et al., submitted for publication). L1, a member of the immunoglobulin superfamily, has been reported to be associated with a number of actions in the nervous system pertaining to important developmental and regenerative processes, including neurite outgrowth and fasciculation, myelination, synaptic plasticity, axonal regeneration, and cellular migration.[35] During these processes, L1 has been shown to be involved in homophilic and heterophilic signal transduction events leading to changes in steady-state levels of intracellular messengers involving, for example, the mitogen-activated protein kinase (MAPK) cascade and regulating gene transcription.[36,37] Thus, the major aim of our study was to gain insight into the contributions of L1 (as a prototypical representative of other adhesion molecules such as CAMS, integrins, cadherins, and others) in helping the bi-directional communication of graft and host elements.

Using the same MPTP-based lesioning paradigm mentioned above and various permutations of L1 overexpression in grafted NSCs and/or host astrocytes (the astrocyte having been identified as a cell type pivotal for defining the stem cell niche), we indeed found beneficial effects of this cell-recognition molecule on donor cell migration and survival, which resulted in faster and more robust recovery of the crippled host DA neurons (Ourednik et al., submitted for publication). Intriguingly, this was obvious even in animals receiving time-delayed (4 weeks post MPTP) grafts. Thus, as originally postulated, in the hierarchy of molecular interactions between SCs and the adult CNS, adhesion/recognition molecules may play a pivotal—perhaps enabling—role in allowing SCs to help restore a homeostatic milieu, and a controlled manipulation of molecules like L1 could, someday, become an additional and important tool in the design of more efficient stem cell-based therapies.

CNS PROTECTION VIA "STEM CELL VACCINATION"

In the preceding sections, we have adduced ample evidence supporting our hypothesis that grafted and residential SCs do collaborate in developmental and regenerative processes of the mammalian CNS and can play decisive roles in the reestablishment of the homeostatic milieu. However, we can speculate even further that by seeding the nervous system with SCs prior to the occurrence of damage we might actually prevent the often devastating sequelae of many hereditary diseases and CNS trauma. We tried to investigate the usefulness of such "preventive" grafting by using spontaneous cerebellar mouse mutants in the following set of experiments.

A number of naturally occurring mouse mutants exist that emulate various aspects of neurodegenerative diseases. Many of these mutants are characterized by the degeneration of Purkinje cells (PCs), an important but quite vulnerable neuronal cell type within the brain. Three mouse mutants that we have begun to use in pilot studies are, in increasing severity and virulence, the *nervous* (*nr*) mouse, the *Purkinje cell degeneration* (pcd) mouse, and the *Lurcher* (*Lc*) mouse. Ostensibly, the only similarity these three mutants have in common is their loss of PCs (and secondarily granule cell neurons) at various times throughout their lives, usually beginning at early postnatal time points; the etiology for PC death seems to vary from mutant to mutant, although further investigation in this regard is ongoing. In all three mutants, however, grafted NSCs, depending on the time of their transfer, appeared to exert an impressive protective influence on the vulnerable PC population. While a small number of NSCs differentiated into PC-like cells, the overall appearance of PC reconstitution derived from the rescue of *host* PCs. This strong potential by NSCs to preserve neuronal populations in the mutant cerebellum seemed most prominent if NSCs were implanted before or at the onset of PC degeneration—that is, on postnatal days 0–10 (P0–10)—rather than during more advanced stages of PC loss (postnatal day 14 to 12 months). This positive effect of the grafts was reflected in both an impressively improved cerebellar cytoarchitecture and in the correction of motor behavior in the host animals.[38]

We have to bear in mind that an important component of many such neurodegenerative disorders—including Parkinson's disease and the PC death in our mouse mutants—is the occurrence of oxidative stress. The latter can be induced in experimental animal models by chemicals like 3-nitropropionic acid (3-NP), a drug that evokes progressive degeneration of striatal neurons and leads to behavioral changes similar to those accompanying Huntington's disease in humans.[39] By using this approach, we are presently testing whether it might actually be the buffering and homeostasis-maintaining capacity of grafted NSCs to diminish the impact of oxidative stress on host cells that allows their preventive usage *prior* to the occurrence of the insult in the CNS. Although preliminary, the first sets of data are very encouraging and indicate that preventive grafting can indeed result in substantial rescue of neurons from oxidative stress by the mobilization of protective antioxidant mechanisms (for more details see the paper by Madhavan *et al.* in this volume).

SUMMARY

The research summarized in this review provides us with rather strong evidence that grafted NSCs can be a rich source of cells as well as factors helping the injured CNS to cope with or recover from an unfavorable environment and protecting its structural integrity and functionality. A better understanding of the regulatory signaling mechanisms and of the molecules involved should lead not only to a better understanding of brain development, but also to a better control of NSC behavior in neurotransplantation, which might help to open new therapeutic avenues. The discovery of neurogenesis and persisting endogenous NSCs even in the adult CNS, and the evidence that these dormant progenitors can be stimulated to neurogenesis by an injury and can integrate functionally into existing adult neural networks, have rendered the scientific community more receptive and positively inclined to accept and further evaluate an increased role of the host in its repair. In consequence, this trend allowed our pioneer studies from the '90s to regain significance and to be revisited.

The effectiveness of neural transplantation depends on, among other factors, the type of NSCs used, as well as on their survival and ability to successfully infiltrate damaged brain regions. Unfortunately, both latter processes are drastically reduced in adult or aged CNS and in the presence of chronic insults. The reasons for this include a lack of neurotrophic factors, the presence of myelin-associated inhibitory molecules, and age- or lesion-dependent accumulation of particular variants of the ECM (e.g., chondroitine sulphate proteoglycans) interfering with the signaling between NSC cells and their environment.[1,2,20,29,30] It seems, therefore, that in order to augment regeneration and plasticity in the adult and aged CNS, we may need a combinatorial approach, which would include cell replacement, trophic support, protection from oxidative stress, and the neutralization of the growth-inhibitory and tissue-plasticity-preventing components with "lubricating" molecules like L1. Such a global approach cannot be overestimated and needs to be our highest priority in achieving the best possible in graft-aided CNS regeneration.

REFERENCES

1. RAO, M.S. & M.P. MATTSON. 2001. Stem cells and aging: expanding the possibilities. Mech. Ageing Dev. **122:** 713–734.
2. HORNER, P.J. & F.H. GAGE. 2002. Regeneration in the adult and aging brain. Arch. Neurol. **59:** 1717–1720.
3. ARNEMANN, J. 1787. Versuche ueber die Regeneration an lebenden Thieren. *In* I. Bd. Ueber die Regeneration de Nerven. J. Ch. Dieterich, Ed. Goettingen.
4. BËLSCHOWSKY, M. 1901. Zur Histologie der Compressionveraenderungen des Rueckenmarks bei Wirbelgeschwuelsten. Neur. Zbl. **20:** 217–221, 242–255, 300–305, 344–353.
5. CAJAL, R. 1928/1959. Degeneration and Regeneration of the Nervous System. May R.M. (Ed. and Transl.). Heiner. New York.
6. THOMPSON, W.G. 1890. Successful brain grafting. N.Y. Med. J. **51:** 701–702.
7. SALTYKOV, S. 1905. Versuche ueber Gehirntransplantation, zugleich ein Beitrag zur Kenntniss der Vorgaenge an den zellulaeren Gehirnelementen. Arch. Psychiatr. Nervenkr. **40:** 329–388.
8. DUNN, E.H. 1917. Primary and secondary findings in a series of attempts to transplant cerebral cortex in the albino rat. J. Comp. Neurol. **27:** 565–582.
9. PERLOW, M.F., W.F. FREED, B.J. HOFFER, *et al.* 1979. Brain grafts reduce motor abnormalities produced by destruction of nigrostriatal dopamine system. Science **204:** 643–647.
10. XIAO BG & H. LINK. 1998. Immune regulation within the central nervous system. J. Neurol. Sci. **157:** 1–12.
11. BRUNDIN, P., M. EMGARD & U. MUNDT-PETERSEN. 1999. Grafts of embryonic dopamine neurons in rodent models of Parkinson's disease. *In* CNS Regeneration: Basic Science and Clinical Advances. Tuszynski M.H. & J.H. Kordower, Eds. :299–320. Academic Press. San Diego, CA.
12. SLADEK, J.R., T.J. COLLIER, J.D. ELSWORTH, *et al.* 1999. Fetal grafts in Parkinson's disease. *In* CNS Regeneration: Basic Science and Clinical Advances. Tuszynski M.H. & J.H. Kordower, Eds. :321–364. Academic Press. San Diego, CA.
13. BJORKLUND, A. & O.V. LINDVALL. 2000. Cell replacement therapies for central nervous system disorders. Nat. Neurosci. **3:** 537–544.
14. BACHOUD-LEVI, A.C., P. HANTRAYE & M. PESCHANSKI. 2002. Fetal neural grafts for Huntington's disease: a prospective view. Mov. Disord. **17:** 439–444.
15. FREEMAN, T.B., A. WILLING, T. ZIGOVA, *et al.* 2001. Neural transplantation in Parkinson's disease. Adv. Neurol. **86:** 435–445.

16. RENFRANZ, P.J., M.G. CUNNINGHAM & R.D. MCKAY. 1991. Region-specific differentiation of the hippocampal stem cell line HiB5 upon implantation into the developing mammalian brain. Cell **66**: 713–729.
17. SNYDER, E.Y., D.L. DEITCHER, C. WALSH, S. ARNOLD-ALDEA, et al. 1992. Multipotent neural cell lines can engraft and participate in development of mouse cerebellum. Cell **68**: 33–51.
18. HIRAI, H. 2002. Stem cells and regenerative medicine. Hum. Cell. **15**: 190–198.
19. OUREDNIK V, J. OUREDNIK, K.I. PARK, et al. 2000. Neural stem cells are uniquely suited for cell replacement and gene therapy in the CNS, Novartis Found. Symp. **231**:242–62.
20. GAGE, F.H., P.W. COATES, T.D. PALMER et al. 1995. Survival and differentiation of adult neuronal progenitor cells transplanted to the adult brain. Proc. Natl. Acad. Sci. USA **92**: 11879–11883.
21. HERRERA, D.G., J.M. GARCIA-VERDUGO & A. ALVAREZ-BUYLLA. 1999. Adult-derived neural progenitors transplanted into multiple regions in the adult brain. Ann. Neurol. **46**: 867–877.
22. FRICKER, R.A., M.K. CARPENTER, C. WINKLER, et al. 1999. Site-specific migration and neuronal differentiation of human neural progenitor cells after transplantation in the adult rat brain. J. Neurosci. **19**: 5990–6005.
23. OUREDNIK, J., W. OUREDNIK & H. VAN DER LOOS. 1993. Do foetal neural grafts induce repair by the injured juvenile neocortex? Neuroreport **5**: 133–136.
24. ROSSINI P.M. & G. DAL FORNO. 2004. Neuronal post-stroke plasticity in the adult. Restor. Neurol. Neurosci. **22**: 193–206.
25. OUREDNIK, J., V. OUREDNIK, W.P. LYNCH, et al. 2002. Neural stem cells display an inherent mechanism for rescuing dysfunctional neurons. Nat. Biotechnol. **20**: 1103–1110.
26. OUREDNIK, J., W. OUREDNIK & D.E. MITCHELL. 1998. Remodeling of lesioned kitten visual cortex after xenotransplantation of fetal mouse neopallium. J. Comp. Neurol. **395**: 91–111.
27. OLSON L, C. AYER-LELIEVRE, T. EBENDAL, et al. 1990. Grafts, growth factors and grafts that make growth factors. Prog. Brain Res. **82**: 55–66.
28. LU, P., L.L. JONES, E.Y. SNYDER, et al. 2003. Neural stem cells constitutively secrete neurotrophic factors and promote extensive host axonal growth after spinal cord injury. Exp. Neurol. **181**: 115–129.
29. YAMASHITA Y.M., M.T. FULLER & D.L. JONES. 2005. Signaling in stem cell niches: lessons from the *Drosophila* germline. J. Cell Sci. **118**: 665–672.
30. GARCION E, A. HALILAGIC, A. FAISSNER & C. FFRENCH-CONSTANT. 2004. Generation of an environmental niche for neural stem cell development by the extracellular matrix molecule tenascin C. Development **131**: 3423–3432.
31. OUREDNIK, V., J. OUREDNIK, J.D. FLAX, et al. 2001. Segregation of human neural stem cells in the developing primate forebrain. Science **293**: 1820–1824.
32. TENG, Y.D., E.B. LAVIK, X. QU, et al. 2002. Functional recovery following traumatic spinal cord injury mediated by a unique polymer scaffold seeded with neural stem cells. Proc. Natl. Acad. Sci. USA **99**: 3024–3029.
33. PARK K.I., Y.D. TENG & E.Y. SNYDER. 2002. The injured brain interacts reciprocally with neural stem cells supported by scaffolds to reconstitute lost tissue. Nat. Biotechnol. **20**: 1111–1117.
34. SCHMID, R.S. & P.F. MANESS. 2001. Cell recognition molecules and disorders of neurodevelopment. *In* International Handbook on Brain and Behavior in Human Development. A.F. Kavelboer & A. Gramsbergen, Eds. :199–218. Kluwer. Groningen. Netherlands.
35. BRUMMENDORF, T., S. KENWRICK & F.G. RATHJEN. 1998. Neural cell recognition molecule L1: from cell biology to human hereditary brain malformations. Curr. Opin. Neurobiol. **8**: 87–97.
36. SCHAEFER, A.W., H. KAMIGUCHI, E.V. WONG, et al. 1999. Activation of the MAPK signal cascade by the neural cell adhesion molecule L1 requires L1 internalization. J. Biol. Chem. **274**: 37965–37973.
37. SILLETTI, S., M. YEBRA, B. PEREZ, et al. 2004. Extracellular signal-regulated kinase (ERK)-dependent gene expression contributes to L1 cell adhesion molecule-dependent motility and invasion. J. Biol. Chem. **279**: 28880–28888.

38. OUREDNIK V, J. OUREDNIK, B. KOSARAS, *et al.* 2001. Nerve cell rescue and behavioral changes induced by grafted clonal neural stem cells in ataxic cerebellar mouse mutants. Am. Soc. Neurosci. Abstr. **27:** 371.3.
39. BEAL, M.F., E. BROUILLET, B. JENKINS *et al.* 1993. Neurochemical and histologic characterization of striatal excitotoxic lesions produced by the mitochondrial toxin 3-nitropropionic acid. J. Neurosci. **13:** 1481–1492.

Grafted Neural Stem Cells Shield the Host Environment from Oxidative Stress

LALITHA MADHAVAN, VÁCLAV OUREDNIK, AND JITKA OUREDNIK

Department of Biomedical Sciences, College of Veterinary Medicine,
Iowa State University, Ames, Iowa 50011, USA

ABSTRACT: Here, we present our preliminary data showing that neural stem cells (NSCs) can prevent the degeneration of striatal neurons when transplanted into the CNS prior to intoxication with 3-nitropropionic acid (3-NP). In the adult CNS, the number of NSCs, a major source of neural cell populations and plasticity-modulating factors, is relatively low if compared to that of the developing brain. This, together with the adult growth-inhibitory environment, limits its regenerative capacity. Our recent observation has shown that grafted NSCs may rescue/protect neurons in the chronically impaired mesostriatal system. On the basis of this study and because we were also intrigued by our recent observations regarding the rescue/protective role of NSCs *in vitro*, we decided to test the hypothesis that grafted NSCs can also be deposited preventively in the CNS (and perhaps join the pool of endogenous NSCs of the intact host brain) for later buffering and maintenance of homeostasis when the host is exposed to oxidative stress.

KEYWORDS: CNS; neuroprotection; regeneration; rescue; 3-NP; striatum; Huntington's disease

THE POTENTIAL OF NSCs TO INDUCE REGENERATION OF THE MAMMALIAN CNS

While the major lines of present neurotransplantation research are still focused on the original challenging task of replacing dead or dysfunctional CNS structures, recent studies indicate that neural grafts have the additional ability to awaken latent regenerative capacities within the host CNS.[1] Thanks to sophisticated techniques in molecular neurobiology, many aspects of this stimulating interaction between graft and host are constantly being unraveled and are providing us with a growing number of factors, receptors, and signaling pathways involved in normal ontogeny of the mammalian CNS and its many pathologies.[2,3]

While donor NSCs engraft into the recipient parenchyma, they become tightly embedded into the host environment and a dialogue ensues between graft and host elements leading to dynamic cellular and molecular changes on both sides. During this process, the rigidity of the more mature, adult, or aged host CNS is shifted back

Address for correspondence: Dr. Jitka Ourednik, Department of Biomedical Sciences, College of Veterinary Medicine, Iowa State University, Ames, IA 50011. Voice and fax: +1-515-294-6449.
joured@iastate.edu

Ann. N.Y. Acad. Sci. 1049: 185–188 (2005). © 2005 New York Academy of Sciences.
doi: 10.1196/annals.1334.017

into more plastic stages, and the donor tissue receives signals to differentiate according to the actual program of the host.

We have demonstrated that while some of the grafted cells mature into various specific neural cell types, a substantial number remains mostly undifferentiated.[1,4] On the basis of further histological and molecular analyses, we proposed that these immature graft cells may—similar to the dormant cells of the host—be important for the observed "rescue" of the chronically impaired host neurons.[1] Specifically, in our previous study,[1] the focus was on the dopaminergic cells in the substantia nigra compromised by aging and superimposed 1-methyl-4-phenyl-1,2,3,6-tetrahydropyridine (MPTP) treatment. Although some of the donor NSCs did contribute to nigral reconstitution, the majority of dopaminergic cells were of host origin, suggesting that they represented rescued dysfunctional cells. The presence of large numbers of undifferentiated progenitors expressing GDNF within and close to the impaired area provided a plausible molecular basis for this finding. Therefore, although injury alone triggers the endogenous release of important plasticity-modulating substances,[5,6] their concentration and effects can be substantially enhanced after transplantation as they can also be produced by the donor[1,7] tissue, resulting in a multiplication in their effects. In summary, these observations suggest that host structures may benefit from not only the NSC-derived replacement of dead or dying cells, but also from the innate capacity of NSCs to create environments rich in neuroprotective substances and support host survival.

GRAFTED NSCs CAN ENHANCE RESISTANCE OF THE ADULT MOUSE BRAIN TO OXIDATIVE STRESS

The aim of the present pilot experiment was to investigate whether grafted NSCs can protect host animals from the behavioral and pathological effects of 3-NP —an oxidative stress–causing agent—when transferred *prior* to the insult.

Mice were grafted with NSCs (derived from the subventricular zone of newborn mice, kind gift of Stefan Krauss, University of Oslo, Norway) one week before repeated intoxication with the mitochondrial toxin 3-NP.[8] Normally, the latter regime induces acute (lasting several days) Huntington's disease-like symptoms reflected in structural changes in striatum and mesencephalon and in movement abnormalities. Behavioral changes were monitored in several tests described previously by Fernagut *et al.*[8] The animals were sacrificed 3 days later and histologic evaluation of the brain cytoarchitecture used to compare control (intact and mock-grafted) and grafted/lesioned groups.

We observed that while all "mock-operated" animals (no graft prior the chemical insult) suffered from acute and severe behavioral changes typical for 3-NP intoxication, mice protected by intra-striatal grafting prior to the chemical treatment expressed no or only mild symptoms. Moreover, the grafted mice recovered substantially faster than the control animals. These outcomes were in harmony with subsequent immunohistochemical evaluations revealing significantly less neurodegeneration in the striata of preventively grafted hosts. The majority of preserved neurons in these brains remained of host origin, suggesting that the integrity of the host cells had been preserved in the face of 3-NP intoxication. Confirming our hypothesis, donor NSCs remained mostly undifferentiated (nestin-positive) and were found to be immunoreactive for ciliary neurotrophic growth factor (CNTF), which has been found to be therapeutically effective in Huntington's disease models.[9,10] This inherent secretion of CNTF (and probably other factors) by grafted cells is likely to have provided trophic and neuroprotective support to the host neurons and to have contributed towards the observed neuroprotective effect.

CONCLUSIONS

Our preliminary data suggest that NSCs have a significant inherent capacity to protect metabolically compromised host neurons from oxidative stress-mediated damage. Neural cells can utilize various defense mechanisms in order to maintain their integrity in face of environmental demands and after CNS injury. These mechanisms are complex and include the production of neurotrophic factors and cytokines, expression of various cell survival molecules such as antioxidant and anti-apoptotic proteins, presence of telomerase and DNA repair proteins to protect the genome, and mobilization of NSCs to replace/repair damaged cells.[11] Many of these components allowing a nervous system to regenerate after injury are readily available, but most brain regions do not normally utilize or mobilize them to a sufficiently high degree. This might be ascribed to both intrinsic properties (genetic predisposition) of CNS neurons and influences of the surrounding environment, such as aging and various forms of external insults. Our present study together with some recent reports[2,7,12] demonstrates that NSCs can be used as a tool to interrogate the host and to "boost" the stimulation of the brain's latent capacity for self-repair and neuroprotection. Thus, the interaction of donor and recipient cells under pathological stress— the bi-directional induction of key signaling molecules and the resultant neuroprotection— may be a route by which dormant constitutive and self-reparative developmental programs from within the host CNS can re-emerge. Understanding the cellular responses that emerge from this bi-directional communication between the NSCs and their environment will be critical to developing any realistic cell therapy and in unlocking the brain's capacity for self-repair.

ACKNOWLEDGMENTS

We thank Heidi Gabel and Nada Pavlovic for technical help. This work was supported by grants from Iowa State University to J.O. and V.O.

REFERENCES

1. OUREDNIK, J., V. OUREDNIK, W.P. LYNCH, *et al.* 2002. Neural stem cells display an inherent mechanism for rescuing dysfunctional neurons. Nat. Biotech. **20:** 1103–1110
2. MADHAVAN, L., V. OUREDNIK & J. OUREDNIK. 2004. Neural stem cells (NSCs) protect primary neural cell cultures from chemically induced oxidative stress. Am. Soc. Neurosci. Abstr. 729.16, San Diego, CA.
3. IMITOLA, J., K.I. PARK & Y.D. TENG. 2004. Stem cells: cross-talk and developmental programs. Philos. Trans. R. Soc. Lond. B. Biol. Sci. **359:** 823–837

4. OUREDNIK, V., J. OUREDNIK, J. FLAX, *et al.* 2001. Segregation of human neural stem cells in the developing primate forebrain. Science **293:** 1820–1824.
5. GALL, C.M. & P.J. ISACSON. 1989. Limbic seizures increase neuronal production of messenger RNA for nerve growth factor. Science **245:** 758–761.
6. HICKS, R.R., V.B. MARTIN & L. ZHANG. 1999 Mild experimental brain injury differentially alters the expression of neurotrophin and neurotrophin receptor mRNAs in the hippocampus. Exp. Neurol. **160:** 469–478.
7. PARK, K.I., Y.D. TENG & E.Y. SNYDER. 2002 The injured brain interacts reciprocally with neural stem cells supported by scaffolds to reconstitute lost tissue. Nat. Biotech. **20:** 1111–1117.
8. FERNAGUT, P.O., E. DIGUET, N. STEFANOVA, *et al.* 2002. Subacute systemic 3-nitropropionic acid intoxication induces a distinct motor disorder in adult C57Bl/6 mice: behavioral and histopathological characterization. Neuroscience **114:** 1005–1017.
9. EMERICH, D.F., S.R. WINN, P.M. HANTRAYE, *et al.* 1997 Protective effect of encapsulated cells producing neurotrophic factor CNTF in a monkey model of Huntington's disease. Nature **386:** 395–399.
10. REGULIER, E., L. PEREIRA DE ALMEIDA, B. SOMMER, *et al.* 2002. Dose-dependent neuroprotective effect of ciliary neurotrophic factor delivered via tetracycline-regulated lentiviral vectors in the quinolinic acid rat model of Huntington's disease. Hum. Gene Ther. **13:** 1981–1990.
11. MATTSON, M.P., W. DUAN, S.L. CHAN, *et al.* 2002. Neuroprotective and neurorestorative signal transduction mechanisms in brain aging: modification by genes, diet and behavior. Neurobiol. Aging **23:** 695–705.
12. RYU, J.K., J. KIM, S.J. CHO & K. HATORI. 2004. Proactive transplantation of human neural stem cells prevents degeneration of striatal neurons in a rat model of Huntington disease. Neurobiol. Dis. **16:** 68–77.

Administration of Allogenic Stem Cells Dosed to Secure Cardiogenesis and Sustained Infarct Repair

ATTA BEHFAR,[a] DENICE M. HODGSON,[a] LEONID V. ZINGMAN,[a] CARMEN PEREZ-TERZIC,[a,b] SATSUKI YAMADA,[a] GARVAN C. KANE,[a] ALEXEY E. ALEKSEEV,[a] MICHEL PUCÉAT,[a,c] AND ANDRE TERZIC[a]

[a]Division of Cardiovascular Diseases, Departments of Medicine, Molecular Pharmacology and Experimental Therapeutics, and [b]Physical Medicine and Rehabilitation, Mayo Clinic College of Medicine, Rochester, Minnesota 55905, USA

[c]Centre de Recherches de Biochimie Macromoléculaire, CNRS FRE 2593, Montpellier, France

ABSTRACT: The mitotic capacity of heart muscle is too limited to fully substitute for cells lost following myocardial infarction. Emerging stem cell–based strategies have been proposed to overcome the self-renewal shortfall of native cardiomyocytes, yet there is limited evidence for their capability to achieve safe de novo cardiogenesis and repair. We present our recent experience in treating long-term, infarcted hearts with embryonic stem cells, a prototype source for allogenic cell therapy. The cardiogenic potential of the engrafted murine embryonic stem cell colony was pre-tested by in vitro differentiation, with derived cells positive for nuclear cardiac transcription factors, sarcomeric proteins and functional excitation-contraction coupling. Eight weeks after infarct, rats were randomized into sham- or embryonic stem cell–treated groups. Acellular sham controls or embryonic stem cells, engineered to express enhanced cyan fluorescent protein (ECFP) under control of the cardiac actin promoter, were injected through a 28-gauge needle at three sites into the peri-infarct zone for serial assessment of functional and structural impact. In contrast to results with sham-treated animals, stem cell therapy yielded, over the 5-month follow-up period, new ECFP-positive cardiomyocytes that integrated with the infarcted myocardium. The stem cell–treated group showed a stable contractile performance benefit with normalization of myocardial architecture post infarction. Transition of embryonic stem cells into cardiomyocytes required host signaling to support cardiac-specific differentiation and could result in tumorigenesis if the stem cell dose exceeded the heart's cardioinductive capacity. Supported by the host environment, proper dosing and administration of embryonic stem cells is thus here shown useful in the chronic management of cardiac injury promoting sustained repair.

KEYWORDS: embryonic stem cells; regeneration; therapy; tumor; teratoma; heart

Address for correspondence: Dr. Andre Terzic, Guggenheim 7F, Mayo Clinic, 200 First Street S.W., Rochester, MN 55905. Voice: 507-284-9199; fax: 507-284-9111.
terzic.andre@mayo.edu

Ann. N.Y. Acad. Sci. 1049: 189–198 (2005). © 2005 New York Academy of Sciences.
doi: 10.1196/annals.1334.018

INTRODUCTION

Despite advances in disease management, myocardial infarction remains a leading cause of morbidity and mortality.[1] Owing to the limited capacity of native cardiomyocytes for self-renewal,[2] the myocardium is particularly vulnerable to coronary occlusion, which precipitates irreversible injury and poor outcome.[3] Current efforts to reduce the extent of myocardial damage relies on available palliative therapeutic modalities that do not adequately repair cardiac scarring or wall thinning, and are unable to reverse progressive organ failure after myocardial infarction.[4] This lack of curative therapy warrants the establishment of approaches capable of effectively rebuilding, post infarction, at least part of the dysfunctional myocardium, securing new contracting heart muscle necessary for salvage of organ function.

With the discovery of cellular populations that maintain the capacity to differentiate into specialized cell types, recent advances in developmental biology have provided the foundation for potential reparative solutions on the basis of cell-based organ regeneration.[5–9] A case in point is the use of embryonic stem cells, recognized for a unique regenerative propensity imparted by unequaled pluripotency and proliferative capacity.[10–15] However, the challenge to developing safe therapy from embryonic stem cells is whether these cells can be appropriately dosed in order to differentiate, integrate and ultimately function effectively when transplanted into diseased myocardium while limiting unwanted side effects, such as dysregulated growth.[16–18] Establishing the requirements for long-term efficacy of embryonic stem cell–based functional repair is thus required before this emerging strategy can translate into reliable cell-based cardiotherapy for the infarcted heart. Here, in a model of myocardial infarction, we provide experimental evidence that controlled administration of embryonic stem cells leads to sustained cardiogenesis within the competent host environment resulting in stable therapeutic benefit and repair.

MATERIAL AND METHODS

Embryonic Stem Cells and Derived Cardiomyocytes

Field-emission scanning electron microscopy was used to visualize murine embryonic stem cells in culture.[14,18] Differentiation *in vitro* was achieved using the hanging-drop method to generate embryoid bodies from which, following enzymatic dissociation, a highly enriched population of cardiomyocytes was isolated with a density gradient protocol.[11,14] Expression of cardiac markers was probed by laser confocal microscopy, with action potential activity captured by the patch-clamp method in the current clamp mode.[18]

Stem Cell Transplantation into Host Heart

To track transplanted cells *in vivo*, embryonic stem cells were engineered to express the enhanced cyan fluorescent protein (ECFP) under control of the cardiac-specific α-actin promoter.[11,19] Using a 28-gauge needle, engineered stem cells were delivered directly into healthy or infarcted left ventricular walls of isoflurane-anesthetized mice or rats, respectively.[18] Functional outcome was monitored by short axis 2-D and M-mode echocardiography or by ventricular pressure recordings using a micropressure catheter.[11,18,20] Four to sixteen weeks post injection, hearts were excised and examined by light and epifluorescent wide-field microscopy for structural assessment and presence of stem cell–derived cardiomyocytes, as well as for myocardial electrical activity using the 12-lead electrocardiographic technique.[18]

Statistics

Values are expressed as mean ± standard error. Embryonic stem cell–treated *versus* sham-treated groups were compared using the Student's *t*-test with a *P* value < 0.05 considered significant. The Wilcoxon log-rank test was used for nonparametric evaluation of randomization.

RESULTS

Cardiogenic Potential of Embryonic Stem Cells in Vitro

Once established in culture, embryonic stem cells have the capacity to differentiate from a pluripotent to a cardiac phenotype (FIGS. 1A and B). This is here demonstrated as derived cells recapitulated typical cardiomyocyte features, including nuclear translocation of cardiac transcription factors, the myocyte enhancer factor 2C (MEF2C) and the homeodomain transcription factor (Nkx2.5) leading to sarcomerogenesis (FIGS. 1B and C) and action potential formation associated with L-type Ca^{2+} channel expression (FIGS. 1C–E). Thus, embryonic stem cells serve as a reliable cell-based source for *de novo* cardiogenesis.

Titrated Stem Cell Delivery Secures Cardiogenesis in Vivo

Delivery of embryonic stem cells, engineered for *in vivo* fluorescence tracking, resulted in incorporation into host heart of new cyan fluorescing cardiac cells within the area of stem cell transplantation (FIGS. 2A and B). Engraftment of stem cell–derived cardiomyocytes was associated with normal heart morphology (FIG. 2A) and function (FIG. 2C, *inset*). A threshold in the capacity of the host heart to accept stem cell implantation was established, above which excessive stem cell load compromised cardiogenesis post transplantation owing to uncontrolled differentiation resulting in tumorigenesis (FIGS. 2C and D). Thus, the host heart has a finite capacity for driving cardiogenesis in support of differentiation and engraftment of stem cells *in vivo* (FIG. 2E).

Infarct Repair with Stem Cell Transplantation Yields Stable Outcome

A myocardial infarct model was generated by coronary ligation, resulting in an aneurysmal and scarred anterior wall with significant compromise in contractile performance (FIGS. 3A and B). To test the efficacy of embryonic stem cell-based regeneration of diseased myocardium, embryonic stem cells engineered to fluoresce upon cardiogenesis were delivered by direct myocardial transplantation into the peri-

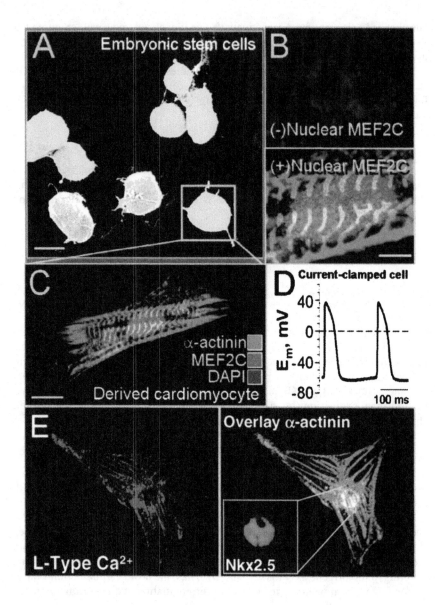

FIGURE 1. Cardiogenic potential of embryonic stem cells. (**A**) Field emission scanning electron microscopy of murine embryonic stem (ES) cells in culture. (**B**) Laser confocal microscopy of an embryonic stem cell (*upper part of panel*) and derived cardiomyoyte (*lower part*) highlighting nuclear translocation of the cardiac transcription factor MEF2C in cardiogenesis. (**C–E**) Cardiac phenotype recapitulated in embryonic stem cell–derived cardiomyocytes including sarcomerogenesis (**C**), action potential activity (**D**), L-type calcium channel expression (**E**), and nuclear localization of the cardiac transcription factor Nkx2.5 (*inset*). Bars correspond to 5, 4, and 15 mm in **A**, **B** and **C**, respectively. Bar in **C** applicable also to **E**.

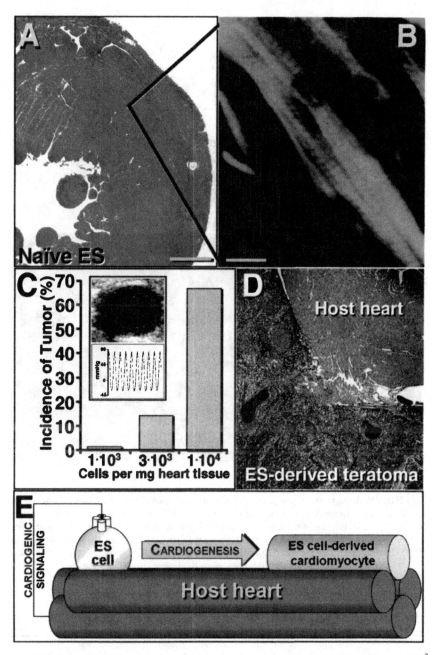

FIGURE 2. Finite capacity of host heart to secure cardiogenesis. (**A**) Delivery of 10^3 naïve embryonic stem (ES) cells/mg of host heart tissue results in proper implantation and differentiation with generation of new cardiomyocytes (**B**). Bars correspond to 2 mm and 12 mm in **A** and **B**, respectively. (**C**) Excessive embryonic stem cell load harbors an increased risk for tumorigenesis. Therapeutic load was established at 10^3 embryonic stem

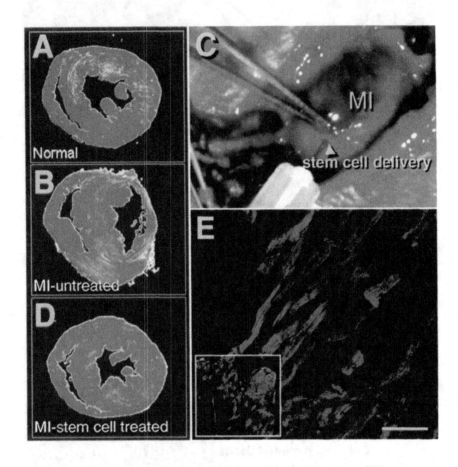

FIGURE 3. Safe stem cell delivery for heart repair. (**A–D**) Comparison of normal (**A**) *versus* infarcted hearts either untreated (**B**) or treated following direct myocardial delivery (**C**) of stem cells (**D**). Note the benefit of stem cell therapy on repair of the anterior left ventricular wall. (**E**) Presence of stem cell–derived cardiomyocytes tracked by expression of the ECFP fluorescent protein under control of the cardiac actin promoter. *Inset* depicts large area of *de novo* cardiogenesis through low-magnification confocal microscopy.

---

cells/mg of host heart tissue, which results in safe implantation with no tumor formation, and sustained normal cardiac structure and function (*inset*). (**D**) Teratoma generated following delivery of an embryonic stem cell overload ($>10^3$/mg host heart tissue). (**E**) Host heart guides embryonic stem cell differentiation with, however, a finite cardiogenic signaling capacity.

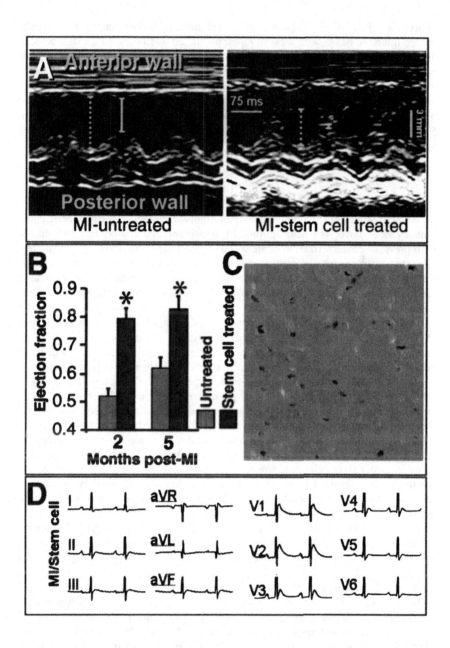

FIGURE 4. Stable benefit of stem cell therapy in myocardial infarction. (**A–B**) Echocardiographic M-mode follow-up study demonstrates that, in contrast to the hypo/akinetic anterior wall of untreated infarcted dilated ventricles, stem cell treatment results into normal anterior wall motion and contractile function sustained throughout the 5-month-long observation period, without evidence for rejection (**C**) or arrhythmia (**D**). *Dashed* and *solid lines* indicate diastolic and systolic left ventricular dimensions, respectively.

infarct area (FIG. 3C). Histopathological examination showed benefit in stem cell–treated infarcted hearts with areas of repair populated by fluorescent cardiac cells indicating stem cell–based *de novo* cardiogenesis (FIGS. 3D and E). In contrast to poorly contracting myocardium in untreated infarcted hearts, stem cell–treated hearts demonstrated synchronous functionally recovered heart muscle on the echocardiographic follow-up (FIG. 4A). Structural repair translated into improvement in the overall ejection fraction in the stem cell–treated group, a benefit maintained over the 5-month observation period, with no evidence for inflammatory infiltrates (FIGS. 4B and C) or arrhythmogenesis (Fig. 4D).

DISCUSSION

We here highlight the stable capacity of embryonic stem cells to generate cardiomyocytes *de novo*, and their aptitude for transplantation and organ repair. In line with previous characterization of stem cell–derived cardiomyocytes,[10,12,14] embryonic stem cells readily differentiated both *in vitro* and *in vivo* demonstrating hallmark features of functional myocardium. Indeed, the clonal and proliferative capacities of embryonic stem cells render this cell type a renewable source for heart engraftment.

In this regard, cardiogenic signaling provided by the host cardiac tissue has been shown to be critical in guiding proper differentiation and implantation of delivered stem cells.[11,21] In particular, the paracrine release of transforming growth factor beta (TGFβ) and bone morphogenic protein (BMP) is essential in instructing pluripotent embryonic stem cells toward cardiogenesis.[11,15] Disrupting the ability of stem cells to decode cardiogenic cues[11] or, as shown here, exceeding the cardiogenic signaling capacity of the host heart through excessive cellular load, compromises the ability of the host heart to guide cardiac differentiation of transplanted cells precipitating uncontrolled growth. To limit unguided differentiation and hone embryonic stem cell plasticity towards the cardiac fate, a threshold cell load was here titrated to achieve proper stem cell engraftment without the risk for tumorigenesis. Thus, the realization that the host heart microenvironment has a finite capacity in directing allogenic stem cell differentiation led to the specification of the stem cell load limit for successful transplantation.

Delivery of embryonic stem cells into a chronic model of myocardial infarction at subthreshold levels for tumorigenesis showed a lasting beneficial effect on the function and structure of the diseased myocardium. Specifically, the impact of such therapy on cardiac contractile performance and ventricular remodeling was associated with documented cardiogenesis in the infarct zone.[18] The reparative response elicited by stem cell transplantation would result in a symbiotic relationship between host heart and transplanted cells, with integration of new heart muscle with host myocardium documented histologically, mechanically, and electrically with no evidence for sites of ectopic activity or rejection. Supported by paracrine cardiogenic stimulation provided by the host heart,[11] stem cell implantation could, in turn, potentially lead to activation of endogenous mechanisms of repair ranging from mobilization of resident heart cells to angiogenesis.[2,22,23] Further characterization of the crosstalk between the host and transplanted stem cells is a necessary step towards the full realization of stem cell–based therapy in the management of chronic heart disease.

ACKNOWLEDGMENTS

A.B. is the recipient of the Mayo Clinic M.D./Ph.D. Training Program funding. This work received support from the National Institutes of Health, American Heart Association, Marriott Foundation, Ted Nash Long Life Foundation, Ralph Wilson Medical Research Foundation, Miami Heart Research Institute, and the Mayo Clinic CR20 Program.

REFERENCES

1. AMERICAN HEART ASSOCIATION. 2003. Cardiovascular diseases. *In* Heart Disease and Stroke Statistics—2004 Update. : 3–9. American Heart Association, Dallas, TX.
2. BELTRAMI, A.P., L. BARLUCCHI, D. TORELLA, *et al.* 2003. Adult cardiac stem cells are multipotent and support myocardial regeneration. Cell **114:** 763–776.
3. TOWBIN, J.A. & N.E. BOWLES. 2002. The failing heart. Nature **415:** 227–233.
4. JESSUP, M. & S. BROZENA. 2003. Heart failure. N. Engl. J. Med. **348:** 2007–2018.
5. ORLIC, D., J.M. HILL & A.E. ARAI. 2002. Stem cells for myocardial regeneration. Circ. Res. **91:** 1092–1102.
6. BRITTEN, M.B., N.D. ABOLMAALI, B. ASSMUS, *et al.* 2003. Infarct remodeling after intracoronary progenitor cell treatment in patients with acute myocardial infarction (TOPCARE-AMI): mechanistic insights from serial contrast-enhanced magnetic resonance imaging. Circulation **108:** 2212–2218.
7. OH, H., S.B. BRADFUTE, T.D. GALLARDO, *et al.* 2003. Cardiac progenitor cells from adult myocardium: homing, differentiation, and fusion after infarction. Proc. Natl. Acad. Sci. USA **100:** 12313–12318.
8. MANGI, A.A., N. NOISEUX, D. KONG, *et al.* 2003. Mesenchymal stem cells modified with Akt prevent remodeling and restore performance of infarcted hearts. Nat. Med. **9:** 1195–1201.
9. THOMPSON, R.B., S.M. EMANI, B.H. DAVIS, *et al.* 2003. Comparison of intracardiac cell transplantation: autologous skeletal myoblasts versus bone marrow cells. Circulation **108:** II264–II271.
10. BOHELER, K.R., J. CZYZ, D. TWEEDIE, *et al.* 2002. Differentiation of pluripotent embryonic stem cells into cardiomyocytes. Circ. Res. **91:** 189–201.
11. BEHFAR, A., L.V. ZINGMAN, D.M. HODGSON, *et al.* 2002. Stem cell differentiation requires a paracrine pathway in the heart. FASEB J. **16:** 1558–1566.
12. SACHINIDIS, A., B.K. FLEISCHMANN, E. KOLOSSOV, *et al.* 2003. Cardiac specific differentiation of mouse embryonic stem cells. Cardiovasc. Res. **58:** 278–291.
13. NIR, S.G., R. DAVID, M. ZARUBA, *et al.* 2003. Human embryonic stem cells for cardiovascular repair. Cardiovasc. Res. **58:** 313–323.
14. PEREZ-TERZIC, C., A. BEHFAR, A. MERY, *et al.* 2003. Structural adaptation of the nuclear pore complex in stem cell-derived cardiomyocytes. Circ. Res. **92:** 444–452.
15. MERY, A., E. PAPADIMOU, D. ZEINEDDINE, *et al.* 2003. Commitment of embryonic stem cells toward a cardiac lineage: molecular mechanisms and evidence for a promising therapeutic approach for heart failure. J. Muscle Res. Cell. Motil. **24:** 269–274.
16. MENASCHE, P. 2004. Embryonic stem cells pace the heart. Nat. Biotechnol. **22:** 1237–1238.
17. KEHAT, I., L. KHIMOVICH, O. CASPI, *et al.* 2004. Electromechanical integration of cardiomyocytes derived from human embryonic stem cells. Nat. Biotechnol. **22:** 1282–1289.
18. HODGSON, D.M., A. BEHFAR, L.V. ZINGMAN, *et al.* 2004. Stable benefit of embryonic stem cell therapy in myocardial infarction. Am. J. Physiol. **287:** H471–H479.
19. MEYER N., M. JACONI, A. LANDOPOULOU, *et al.* 2000. A fluorescent reporter gene as a marker for ventricular specification in ES-derived cardiac cells. FEBS Lett. **478:** 151–158.
20. O'COCHLAIN, D.F., C. PEREZ-TERZIC, S. REYES, *et al.* 2004. Transgenic overexpression of human DMPK accumulates into hypertrophic cardiomyopathy, myotonic myopathy and hypotension traits of myotonic dystrophy. Hum. Mol. Genet. **13:** 2505–2518.

21. OLSON, E.N. & M.D. SCHNEIDER. 2003. Sizing up the heart: development redux in disease. Genes Dev. **17:** 1937–1956.
22. SCHUSTER, M.D., A.A. KOCHER, T. SEKI, *et al.* 2004. Myocardial neovascularization by bone marrow angioblasts results in cardiomyocyte regeneration. Am. J. Physiol. **287:** H525–H532.
23. FOLEY, A., & M. MERCOLA. 2004. Heart induction: embryology to cardiomyocyte regeneration. Trends Cardiovasc. Med. **14:** 121–125.

Index of Contributors